PASS

MOT

FIRST TIME

In the same series

What to Watch out for when Buying a Used Car

Uniform with this book

PASS YOUR
MOT
FIRST TIME

by

Christopher James BA, MIMI, AIRTE

PAPERFRONTS

Typeset in 10pt Times Roman by One and a Half Graphics, Redhill, Surrey. Printed and bound in Great Britain by Clays Ltd., Bungay, Suffolk.

The *Paperfronts* and *Right Way* series are both published by Elliot Right Way Books, Brighton Road, Lower Kingswood, Tadworth, Surrey, KT20 6TD, U.K.

To my wife Esther,
my two daughters Doreen and Velma,
and my son James

CONTENTS

ILLUSTRATIONS

INTRODUCTION

"Vehicle testers are sadists."
*"Coming to an MoT Test station is like visiting the dentist
to have one's teeth extracted."*
*"You are thorough — I have never seen
a Test done so painstakingly."*

When you listen to such comments, you do not need to be
a genius to deduce that owners are often very uneasy when they
present their means of transport to be tested!

I have been testing cars and light goods vehicles for almost
11 years and make a conservative estimate of having tested
about 20,000 vehicles during this period. Within this time I
have acquired a wealth of information and experience which I
have put together here, both for the professional tester, and for
the D-I-Y tester wishing to establish in advance that his car or
light goods vehicle should pass.

The aims of my MoT Test manual are to be simple, free of
jargon, and to require no prior mechanical knowledge on the
part of the reader. It is my unflinching belief that you do not
need to be a mechanic to discover whether your car is
roadworthy, once you have understood how to check the
vehicle and to assess wear and tear in its components.

Thus, with this book, anyone ought to be able to examine his
own vehicle meticulously before presenting it for a Test.
Things requiring repair can be identified and dealt with
beforehand. That vehicle should then pass first shot and save
both time and money. (At the time of writing, a huge near 50%
of MoT Tests are being failed.)

Incidentally, if when you go to buy a second-hand vehicle,

you carry out as many of the tests contained in this book as may be feasible at the kerbside or garage dealer, etc., you may well save yourself a lot of expense and trouble. The basic faults you learn to uncover here are the very ones you want to avoid buying!

I hope that the professional tester, no less than the vehicle owner, will welcome my straightforward text.

Chapter 1 describes all the MoT Test checks, taking them in logical sequence and on the same lines as are laid down for them to be carried out under workshop conditions. It should therefore serve as a useful everyday reference to the necessarily detailed and demanding requirements which have quite rightly been developed by the Department of Transport over the years since the MoT Test began.

I have placed particular emphasis on types of fault that can easily remain hidden from the less experienced tester, both to demonstrate the degree of vigilance required to do the job well from a professional point of view, and to alert vehicle owners everywhere about dangerous faults that can still slip through the MoT Test net − despite the carefully structured system under which Tests are carried out.

All experts in motor-vehicle testing seem to agree that the most difficult part of the MoT Test to interpret properly is the assessment of corrosion of a vehicle's structure. It is by its very nature a grey area. What actual degree of degradation must constitute a failure item, and what a pass, is hard to define in absolute terms. The Department of Transport gives appropriate guidelines for different vehicle parts, but, other than for people who have spent a considerable number of years in the trade, these can still be difficult matters upon which to make certain judgment. In this book I have therefore dipped into my years of practical testing in order to simplify for you what are reasonable and correct pass and fail corrosion lines to draw.

Chapter 1 takes you through the MoT Test in **12 Steps**. Chapter 2 and the subsequent chapters I hope will prove invaluable to readers needing to know if they can do particular jobs themselves, or if the work involved would be beyond them, necessitating going to a garage. Wherever I think a job is likely to have to be one for a garage I have said so; where D-I-Y work may be appropriate I have given details — usually by referring to **Part Two** of the book which is in the form of an A-Z of D-I-Y repairs.

THE MoT TEST REQUIREMENTS

The components checked come under six main headings of what are known as **DIRECT TEST ITEMS**. Everything must be in working order.

Lighting equipment function
1) Front lamps (the two front sidelamps)
2) Rear lamps (the two rear sidelamps)
3) Headlamps
4) Headlamps' aim
5) Stop lamps
6) Rear reflectors
7) Direction indicators

Steering suspension
8) Steering controls
9) Steering mechanism
10) Power steering
11) Front-wheel-drive transmission shafts
12) Stub-axle assemblies (to which the wheels attach)
13) Wheel bearings
14) Suspension parts
15) Shock absorbers

Brakes
16) Service brake condition (footbrake)
17) Parking brake condition (handbrake)
18) Service brake performance
19) Parking brake performance
20) Service brake balance (do they pull one way or the other?)

Tyres and wheels
21) Tyre type
22) Tyre condition
23) Road wheels

Seat-belts
24) Security of mountings
25) Condition
26) Operation

General items' function
27) Windscreen washers
28) Windscreen wipers
29) Horn
30) Exhaust system
31) Exhaust silencer
32) Exhaust emission levels
33) Vehicle structure (dangerous corrosion, splitting of
    load bearing members, etc.)

Technically, a tester can fail a vehicle even if only *one* of the above-listed items is not in order. For example, he is empowered to fail a vehicle despite the only thing wrong being that the windscreen washer jets aim too high or too low. However, most testers would endeavour to rectify such a minor fault and then give a pass certificate.

Sometimes wear in a component is marginal or a fault is as yet of little consequence. Then testers are allowed some discretion. However, their main consideration will always be whether the vehicle is likely to become unroadworthy and a danger to other road users if it is driven again before repairs are carried out. If a tester decides to pass a vehicle despite something which will soon need repair, he will draw the motorist's attention to the problem and advise him or her to have the item in question dealt with as soon as possible. For example, a tester might pass a newly-fitted exhaust pipe leaking slightly, say, at the exhaust manifold down-pipe flange (where the exhaust pipe attaches to the exhaust manifold) but fail a similar leak in an ageing exhaust because, whereas the new one can coke up and stop leaking eventually, the leak on the old one can only get worse.

Beyond the Direct Test Items above, are what are known as **INDIRECT TEST ITEMS**. These are items on which a tester can still fail a vehicle, despite a clean bill of health otherwise, most usually on the grounds that they represent some danger to road safety. They can be, for example, problems that prevent the tester from completing the Test safely, or faults which he considers to be dangerous when the vehicle is driven. They can relate to any component, mechanical or structural, not merely those things which happen to be embodied under MoT requirements.

The Department of Transport includes the concept of Indirect Test Items to make certain testers take a good look at *the whole vehicle*. The purpose is to ensure that any defect which may be *potentially dangerous* to road safety, whether on a testable item or not, will at least be notified to the motorist — even if the vehicle can still be passed because the fault has not yet rendered it unsafe.

Here are two detailed examples:

In the first one the universal joints at each end of the customer's propeller shaft are worn to the extent that, if the tester conducts a brake test on his roller brake tester, there is a good chance of the joints disintegrating and the propeller shaft dropping to the floor. The tester would fail the vehicle because in his judgment the severe wear identified prevents him from completing the Test, even though the vehicle may still be safe to drive in terms of Direct Test Items.

Should trying to do the Test result in any damage to the vehicle and lead to some dispute, the tester can also find himself in an invidious position! He must therefore retain the right to use his professional judgment to fail a vehicle in such an instance. He will not do so lightly. In the event of a Department of Transport inspector being called in to any argument, he knows he would have to be able to show a good technical defence for his decision.

In my second example, the tester finds nothing wrong except that the driver's seat is insecure. It is floating fore and aft. He may decide that he can still pass the vehicle, but that he must also warn the motorist that repair needs to be made at the first available moment. (That he would do on the MoT inspection report form.) Or, he may judge that the fault is too far gone and that he must fail the vehicle. He would do so in this example if he considered that the problem had become *dangerous* and liable to prevent the driver from being able to push the brake pedal effectively in an emergency. Equally, he would do it if he felt there was any risk of damage in completing the roller brake tester part of the MoT Test.

Be advised that for any pass issued with a *warning* you are, because it is in writing, on notice to have the repair done promptly; otherwise any insurance claim in an accident could be affected. You could hardly deny knowledge if it had been made clear to you in writing. Driving an unroadworthy vehicle can also lead to serious charges in Court.

Here are some more, Indirect Test Items commonly leading to failure, mainly because the dangers they threaten are immediate and certain:

1) Engine mountings sheared.
2) Gearbox mountings broken.
3) Fuel leak (threatening skiddy roads for others and fire risk).
4) No oil showing on engine dipstick (likelihood of engine seizure).
5) Underframe caked in mud to the extent inspection is impossible.
6) Body panel, door, bumper, etc., about to fall off (could cause major accident).

The cataclysmic results from some of these can be imagined but, as an example, suppose a gearbox drops down and, in turn, the propeller shaft universal joint attaching thereto is shattered. If the forward end of the propeller shaft were then to dig itself into the road, the rear end of the vehicle could be lifted, with unsavoury consequences.

Some examples of Indirect Test Items that the tester may not fail a vehicle on, but on which he *must* still notify the motorist, might be these — where imminent danger is less, but clearly the motorist has a duty to have the matter attended to:

1) Loose-fitting wheel embellisher.
2) Seriously corroded (or split) non load bearing body member, where jagged protruding edges could, or potentially might, endanger the public. (Examples could be wheel arches, door sills, etc.)
3) Door badly rusted around its hinges and destined soon to fall open.
4) Any fitment on the vehicle in such poor condition as to present a threat to road safety.

At the time of writing, further items for inclusion into the MoT Test have been suggested: cracked windscreens, broken mirrors and faulty number plates are among them.

LEGALITIES

The main rules presently in force are these:

a) All cars must pass the MoT Test at or before their 3rd "birthday" on the road.

b) Thereafter a Test must be passed at 12 monthly intervals, again by the exact day. The timing is convenient because it usually coincides with renewal of the road tax fund licence, the application for which requires an MoT pass certificate to be attached. Special rules allow you to have your MoT Test done a week or two earlier in the second and subsequent years, so that you can carry out any repairs in good time if the vehicle fails the Test before the existing certificate runs out. You have to take the old certificate to the test station, for them to allow the new pass certificate then to run for one year less a day from the expiry date of the existing pass certificate.

c) A police officer who stops you on the road can require to see a current MoT Test pass certificate, or for you to produce one within 7 days at any police station you nominate.

d) If you have lost your certificate, the test station or the Department of Transport can produce a duplicate for a small charge.

e) Circumstances in which you can drive your car without a valid test certificate when it ought to have one are limited to specific journeys made in order to get repairs carried out by prior arrangement, and to get to or from a test station for a test appointment. Despite the obvious need for these exemptions you can still be prosecuted if the car is obviously unroadworthy and should be on a transporter, and you should check that your insurance cover would not be invalidated even though the distance may only be a few hundred metres!

f) Should your car fail, the test fee will not be charged again provided you leave the car there and allow the test station concerned to do the necessary repairs and re-test it for you.

   If you take the car away for the time being, then a full test fee will be charged when you return, either to the same test station or to any other one. You should be aware of e) above to the extent that driving home may or may not be legal and that, once your old certificate goes out-of-time, there must be a previously-booked MoT Test appointment before you can then drive to the garage who will do it.

g) You can appeal if you think your car ought not to have failed. I'm afraid the procedure is somewhat laborious and you must not have repairs made before an appeal re-test is done. For details, refer immediately to the Department of Transport's local office.

PART ONE

1

HOW THE TEST IS DONE –
IN 12 STEPS

Whether a roller brake test is carried out before or after the lighting system is checked depends upon the working methods of the tester. The Department of Transport allows flexibility, although they do insist that the underside of a vehicle should be inspected before any brake test is carried out. Specific inspection procedures for individual testable items are laid down wherever necessary.

The 12 Steps of this chapter assume a righthand drive vehicle and take you round it in my own time-saving order.

On straightforward items all you need to know is here. On the more sophisticated checks, for example where specialised equipment is used, you are also referred to Chapter 2 and to later chapters, where necessary detail is explained. Always look there to make sure you capture all the information this book can provide for you.

A variety of exemptions apply to rare, old cars and some home-built specials; however, I cannot cover them all in this short book. I take for granted that, for example, the normal height and spacing requirements for lights have been followed by manufacturers generally, and that your vehicle is of normal specification.

WARNING: *As your engine will be running for part of the time, do not work in an enclosed space and risk inhalation of dangerous exhaust gases.*

Step 1: Stand your vehicle on level ground, facing squarely to a wall one-and-a-half metres away. Switch off the engine. Open your driver's door and examine your seat-belt *webbing* for fraying, especially at the edges, near fixings and anywhere it is likely to rub. Slowly pull out that part of the belt which retracts into the inertia-reel (if fitted) so that you can carefully inspect the webbing there too.

The inertia-reel must be able to retract the belt properly so as to hold it snug around you, and it must lock in emergency. To check this *re-winding* function, pull the belt fully out (slowly); then let go, allowing it to slide back through your fingers. The belt should retract on its own without hindrance. If it won't, you can give gentle assistance; however, if that fails to make it retract, then the inertia-reel re-wind is faulty and you need to fit a replacement belt. To test the inertia-reel *locking* function, pull *very sharply* on the retractable side of the belt. The belt should lock, and hold, until you let go. Having locked, it *must* also release easily as soon as you take the pressure off. Check this. Otherwise the occupant could be trapped after a crash. (Note: the *locking* test is not possible on all models. Some will only lock under deceleration on the road.)

Now look at the anchorage points on both halves of your belt. The fixings must be tight, with bolts, etc., secure. The load bearing area of bodywork through which they attach must be sound, being free of rust, splits, or distortions. Pull hard on the belt so as to detect any weakness or untoward movement at each anchorage. Scrutinize inertia-reel mountings so far as possible, in the same way. If your belt is integral with the car seat, as some are, or partly are, you must include the security of the seat mounting and adjustment mechanism(s) too.

Do up the two halves of the belt at the buckle and pull hard on the belt to test whether this coupling might, without being released, burst open under pressure. If it does, it must either be replaced or repaired. It is also faulty if, when the release tab is pressed, there is any undue difficulty in taking the belt apart. For further information on seat-belts, see **Part Two**.

Next, sit in the driver's seat, reach for the other belt, and repeat the aforementioned procedure. (If more than one passenger front seat-belt is fitted, strictly speaking only that belt next to the nearside passenger door falls under the testing rules. However, it is folly to have any out of order seat-belt and a tester would certainly inform you of it in writing.)

Back seat-belts come next. Check them exactly in all aspects as for the front ones.

If a baby carrier or a carry cot is fitted, the law says the tester must ensure it is properly secured and must fail the vehicle if it is not. However, if the carry cot or baby carrier is removed before the Test, the tester need not be concerned.

With regard to adult rear seat-belts, the law says if the vehicle is fitted with them, then they should be tested. If the belts are removed from a model known to have been fitted with seat-belts before presenting it for a Test, the tester would fail the vehicle.

Step 2: Go back to your driving seat and test your handbrake for excess travel at the lever. Pull it on and release it fully several times whilst you look for undue wear. (For safety, place the vehicle temporarily in gear or make certain automatic transmission is set in Park.) Excessive or insufficient travel are both failure points, though excessive travel is a more important defect because it may prevent setting the handbrake properly. Insufficient travel, provided the brakes do not drag (that is, remain slightly on all the time), may be passed, though a good tester will advise you to have it adjusted.

Inspect the anchorage of the lever. Rust may have weakened the security of the fixings or the panelling on which they are mounted. A minimum area of 30cm surrounding the base must be in solid, unbending, fracture-free material. Check that the ratchet teeth engage fully with the pawl to hold the handbrake hard on. Ensure that a casual knock on the release button or pushing the lever sideways, as might happen if a child was playing or if the driver or a passenger were to climb across to swap front seats, cannot cause accidental release. Also look for sideways slackness when the lever is off. It is then that bad wear at the pivot may most show.

Now examine the footbrake pedal for damage and proper security. If the anti-slip pedal rubber is badly worn or missing, it should be replaced immediately. It is a small but significant part of the braking system, and your vehicle will fail the Test if this rubber is worn smooth, or missing! Your foot can easily slip off such a pedal and cause an accident.

Grasp the footbrake pedal and see if you can move it significantly *sideways*. Almost no sideways wobble whatever ought to be detectable. If there is more than a minuscule amount, the matter must be put right, or your car will fail its Test.

If your vehicle is equipped with power-assisted brakes (consult your owners' handbook if you are not sure), now is the time to test this brake servo, as it is also known. First, pump the brake pedal several times — say 10 to 15 times — to exhaust the vacuum from the rear half (the pedal side) of the vacuum chamber; then, at the end of the last stroke, maintain *light* pressure on the pedal and start the engine *immediately*. If the servo system is serviceable, then as soon as your engine runs, your brake pedal should go down measurably further under this light pressure. If there is no change, the servo unit is inoperative. Stop your engine. Servo repair needs garage equipment and expertise beyond D-I-Y capability, but you may wish to tackle straightforward checks. (See **Part Two**.)

Move next to a creep test on your footbrake, to determine whether there is any leakage in the hydraulic system. Place firm (but not emergency level) pressure on the brake pedal for a minimum of one minute, and preferably for two or three. If the pedal creeps (goes down gently) under this pressure, then either air has got into the system, or brake-fluid is leaking out somewhere or, very often, both. Either fault is a failure item. Another possibility is that the master cylinder (fig. 1) is not holding pressure internally; with this latter fault no external leak would show. The longer you make this creep test the better, because the rate of such creep is dependent on the degree of any leakage. Whereas a copious leak might be obvious straightaway, on too brief a test a very slight leak at one wheel cylinder may produce hardly discernible creep. It could therefore remain hidden.

Another indication of leakage is if pedal travel, before solid resistance is felt, is more than the normal small fraction of total travel. Although this can also be caused by brake linings nearing the end of their days, or other defects, usually a pedal which meets no resistance until it has almost reached the floor shows that your brake hydraulics are in a dangerous state. Spongy resistance, especially if it becomes less with repeated applications, is a sign of air in the system. Air generally gains entry via a leak. Both problems mean your vehicle must fail.

Other places to watch for leaks are at the brake master cylinder and, if applicable, its servo connections; in (fixed) brake piping; in (flexible) brake hoses; and at all joints and unions. See fig. 1. Later on, whilst you make a visual examination of as much of the hydraulic system as is possible without dismantling, you must have an assistant maintain pressure on the brake pedal all the time — *so as to encourage any leaks to show*. This provides a second, and extended, opportunity for any creep at the pedal to be noticed. For your own peace of mind I describe an even more reliable test which you can do separately, at the beginning of Chapter 6, **Part Two**.

Malfunctioning brakes, unless you are expert yourself, require garage repair.

You now turn your attention from the brakes to make two checks from inside the car on the steering. First, see that the steering column outer casing is mounted under the dashboard immovably. (Begin by making sure that if a steering column adjuster is fitted — for different drivers to alter the position of the column — it is locked.) Like seat-belt anchorages, there must be no movement at any mounting point, or in the body part(s) through which they attach.

Next, look at the mechanism of the column itself. When you turn the steering wheel there should be no discernible loss of that movement between the steering wheel and the steering

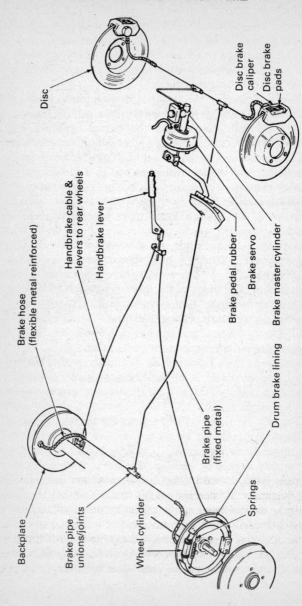

Fig. 1. Hydraulic brake system.

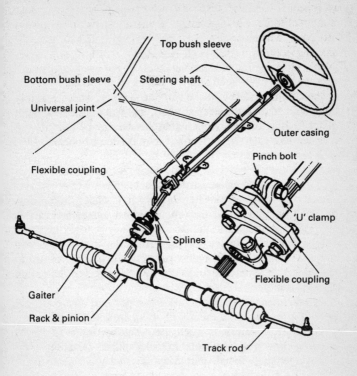

Fig. 2. Steering assembly.

column shaft, caused by, say, looseness of the steering wheel attachment to that steering column shaft. Nor should any play really be evident between the steering column shaft and its own outer casing. To check the latter, grasp the steering wheel and, without turning it, push and pull it firmly in various directions, all at right angles to the column — left and right, top and bottom, etc. Anything more than a touch of play here (seek

expert advice if not sure) should be regarded as excessive. This
is especially so if, when you are turning the wheel, your
rotational force is being reduced significantly before it reaches
the road wheels. Such play here would indicate that the
steering column bushes need renewal.

Still investigating the steering column, you need to push and
pull the steering wheel *axially* (that is, in and out in the line
of the column itself) to make sure the steering wheel is firmly
held, splined onto the steering shaft, and that the steering shaft
is not floating up and down within the steering column. If you
can feel any such "end float", it may be due to normal give,
in a flexible coupling. If so, you must identify that coupling
and make sure nothing is obviously loose, frayed or ill-
fitting, or that the cause is not, in fact, undue wear. See also
Chapter 2. To do so, you may have to look under the bonnet,
or even from underneath the vehicle, whilst an assistant carries
out your instructions at the steering wheel. A torch and/or a
mirror may help you see properly. However, in that case, you
can await **Steps 7** and **9** below respectively, where I return to
the subject. Where these joints are found depends on the design
of your vehicle.

Apart from normal flexing, you can presume that end float
beyond a minuscule amount will result in Test failure.

The biggest culprit is often the pinch-bolt of a U-shaped
clamp getting loose. This clamp locks splined joints together.
The splined male/female parts have serrated edges designed to
interlock in whatever position they are clamped. Very
frequently, all it needs to cure a little end float is to tighten up
the offending pinch-bolt. However, if there are badly chewed
splines caused by a pinch-bolt having been loose for some
time, replacement of major parts may be required.

Many steering columns pass through an angle, or even
several angles. This is facilitated by means of a universal joint
where the column goes "round the corner". While your
assistant turns the steering wheel gently both ways, you need
to watch that the movement above a universal joint is (*as it must
be*) immediately transmitted to the next section down. Unless
that is the case — with no visible looseness in the joint — the

joint is worn and requires replacement.

For the most part you will find that major steering column faults are beyond D-I-Y repair. Fig. 2 shows a typical steering assembly to help you identify parts. See also figs. 19, 20 and 21.

Checking the steering column part of the steering may reveal insignificant wear in itself. However, steering wear tends to be additive, and if there is wear elsewhere in the steering joints too, it may all add up to what we call *lost motion*. That is, instead of the turning of the steering wheel causing a direct instant reaction at the road wheels, an appreciable lagging behind in the movement occurs. You are therefore, not only concerned with the soundness of individual steering components, you must be on the lookout for unacceptable slackness or lost motion in the total system.

The guidelines laid down by the Department of Transport for acceptable (total) lost motion are as follows (the measurement is at the rim of the steering wheel and consists of the amount that you may turn it before movement is seen at the road wheels): (1) if the vehicle has a steering gearbox, up to 7½cm; (2) if rack and pinion steering is fitted, up to 12mm. Figs. 2 and 19 shows a rack and pinion system, as the vast majority of small vehicles now employ this superior type. Fig. 20 shows a steering gearbox.

Although it is very uncommon nowadays, you should note that a steering wheel which has been bent out of true, or dangerously fractured at any point, would have to be failed.

Move on to headlamps. Switch them on and check both high and dipped beams on the wall to see that they are working on both sides of your vehicle. (On a four-headlamp configuration, the outer pair must provide the dipped beam.) Your high beam has to correspond with its tell-tale indicator on the instrument panel, which must also work. The brightness of your headlamps will be considered in **Step 4**; beam-setting will be detailed later, in **Part 2**. Your headlamps' responses to the on/off switch and to the dipswitch must be instant.

If a lamp (or one beam of it) has failed, you should find it simple to fit a replacement, having consulted your owners' handbook. Sometimes an earthing fault is the cause of a complete headlamp failure. See **Part Two,** page 133.

Test your horn. There should be an uninterrupted sound of normal loudness for your model. Intermittent or low volume functioning must be corrected. This usually requires expert attention unless it is a simple earthing fault. See **Part Two**, page 133.

Still in the driving seat, test your windscreen washers and wipers. Observe the area swept by your wipers. If any part is being missed, bear this in mind when you inspect the rubber blades in **Step 3** below. Technically, your washers are required to be able to spray both in sufficient quantity, and in the right place for the most effective screen cleaning. Are they aimed efficiently on both sides? Do they eject cleaner copiously enough? Is that cleaner frost-protected? Remember that not having a clean windscreen is a traffic offence. Although MoT regulations about the windscreen washers mainly refer to the driver's side, it is arguable whether adequate vision for safe driving can be had if a passenger-side washer is inadequate. (Strangely, though the rules require both wipers (or a pair) to function, they demand only that the driver's side washer functions.) The area swept on the passenger-side affords a vital view across towards the nearside pavement. Therefore, unless a car is designed for a single wiper and washer to be sufficient, a tester will probably fail it if the passenger-side washer is below standard. See **Part Two** for mending windscreen washers.

Step 3: You need an assistant again − inside the car − as you move round, taking in various lighting and other checks. From in front of the vehicle, inspect both sidelamps. They must be white lights, come on at normal unflickering brilliance, and have outer lenses that are intact. Where sidelamps are integral with headlamps they must be capable of coming on separately. Rock your headlamps' glass to make sure they are secure in their recesses.

Whilst at the front, look to see whether *both* front tyres are marked radial, or whether they are of the cross-ply type. By law, they must match. (Detailed examination of each tyre comes later; only concern yourself with this fundamental requirement for the moment.)

Next, you test both front indicators. Have your assistant ensure that the indicator "tell-tale" on the instrument panel is either seen or both seen and heard operating in each case, while you check that flashing speeds and brilliance are normal. (No tell-tale is required if the indicators are easy to see from your driving seat.) The law requires indicators to flash at between one and two times per second. If they don't, or the lamps are dim, your vehicle will fail. (A too-low rate may improve and be adequate with the engine running, which is acceptable.) Make certain both your front indicator lenses are either white or amber. No other colours are allowed. Lenses must not be broken, and each lens and lamp as a whole must be securely fixed. Additional indicator lamps provided on the wings of some cars are not always included in the Test though most testers do include them. (**Part Two** may help you mend common indicator malfunctions.)

Next, lean across the windscreen from the driver's side and inspect the nearest wiper blade if there is more than one, or both blades if a nearside one is also accessible. (Combined use of the ignition and wiper switches may enable you to park the blade(s) conveniently.) Lift each blade off the windscreen and run your thumb nail along the base length of the rubber to detect any significant cuts. Look for any breaks in the wiping edge. Check that the wiper arm is reasonably pressed against the windscreen (by spring tension), and that the arm-securing splines are neither worn nor loose. (Fig. 2 shows splines if you are not familiar with them.) Any of these faults, especially a torn rubber blade, is a failure point. (For minor repairs, see **Part Two.**)

Move on to the rear of your vehicle. On your way, examine the outside walls of your two offside tyres. Apart from trivial marks on the outer surface, evidence of splitting or of cuts which reach down to the corded inner walls, or of bulges beginning to appear, have to mean failure. (Ignore the treads at this stage; a thorough examination of these comes at **Step 11** below.) But do inspect the wheel and its rim for distortion. Rims with slightly bent edges can be judiciously hammered back to shape. If this remedial action is unsuccessful, then the wheel has to be expertly repaired, or renewed. A wheel clearly so distorted as to affect the safe steering/driving of the vehicle will be failed. So will one with a rim unlikely to hold the tyre satisfactorily in place at speed. A key point to note is that, with a tubeless tyre, the air seal depends on the rim being sound. Take advice from a tyre specialist if you are in doubt.

Now at the back of your vehicle, deal with your lights. Both rear red sidelamps must come on clearly and not flicker. With your *rear sidelamps still switched on*, call for your stop lights. Look for any *interaction* between the sidelamps and the stop lights (for example, a sidelamp going off as the stop lights come on). This common fault is normally caused by a malfunctioning bulb or a bad earth. Reference to your owners' handbook and **Part Two** of this book may help you rectify the matter. Both your stop lights must be noticeably brighter than your sidelamps, as well as equal to each other. Their on/off responses must accurately match the use of your brake pedal. Check that the lights remain on at a steady brilliance whilst the brake pedal is held on for a reasonable length of time. The outer plastic lenses of the sidelamps and stop lights are often clustered together and include a pair of red reflectors, though the latter may be separately mounted. All these lenses must be intact, secure, and free from discoloration. Your number-plate lamp should work.

Apply to both rear indicators, in turn, the same checks you made on the front indicators.

Still at the back, find out if the rear tyres are properly matched *and* suitable in combination with the front ones. The law says a vehicle should be fitted either with cross-ply tyres all round, or radial-ply tyres all round. It will allow mixing but ONLY if both the cross-ply tyres are at the front, and the pair of radials are at the rear. Each front tyre must match the other. Each rear tyre must match its mate. Mixing of types front to back is banned. Note that there are two basic types of radial tyre. Although legally they can be mixed, it is advisable (to obtain the best traction) to fit either steel-braced radials all round or textile-braced radials all round.

On your way round to the front again, investigate the two nearside wheels and their tyre walls, exactly as you did the offside ones. Also check out your nearside wiper blade if you were unable to reach it before.

Step 4: Return to your headlamps. This time you are concerned with brightness. At the one-and-a-half metres from a wall which you stood your car to begin with, each headlamp should throw a distinct white bright pool of light against the wall, both on dipped and high (or main) beams. (The yellow tinge built into many continental headlamps is permissible on matched pairs of headlamps.) A lamp that emits a dull, brownish-white light on either beam will fail. A bad earth, or a perished reflector in the lamp unit, accounts for practically all cases of dimness. In **Part Two** I investigate bad earths. The root cause of a tarnished reflector is usually condensation inside the lamp. Water and air get in through a cracked or holed front glass to cause the damage and, ultimately, the whole unit needs replacement.

A headlamp with a minor fault in its front glass may still pass the Test provided it produces satisfactory light. The glass must remain secure when pressed gently with the thumb. If cracked glass wobbles, or is obviously about to disintegrate, it must be failed. If there is a small hole but the outdoor elements don't seem to be entering the headlamp, then this too may be passed. However, a headlamp with a larger hole (particularly one that seems to be the cause of condensation affecting it, and with deterioration of the reflector already in evidence) would be failed. The above comments should be your guide as to how a tester is likely to use his judgment. Your best bet is to replace a suspect headlamp.

Your next step is to check out your headlamps' aim (focus). A test station uses a beam-setting machine for this job. Details of how you should be able to carry out your own approximate check, and make any necessary adjustments, are in **Part 2**. Please refer there. Alternatively, you can have a garage complete this task for a small fee.

Step 5: Open your bonnet.

(WARNING: *Keep your fingers, face, clothing, etc., away from possibly hot parts such as the radiator or exhaust system. Beware also of an electric cooling fan. On some models they activate – without warning –* **AFTER** *YOUR ENGINE HAS BEEN SWITCHED OFF.* Usually, a message-plate on top of the radiator "shouts" a message about this to remind you. Tuck a tie well into your shirt – out of harm's way!)

Ask your assistant to press the footbrake hard on, and keep his foot there as hard as he can, maintaining that constant pressure. (If your vehicle has servo-assisted brakes, the engine must be running.) Whilst he does so, you inspect all the brake pipes and unions visible in the under-bonnet area, as well as the servo unit connections (if fitted) and the brake master cylinder. As mentioned earlier, he must report to you if he detects any creep at the brake pedal during this extended inspection. Refer to fig. 1 to help you track all the brake piping in your car. Refer to Chapter 2 to discover exactly what you must look for in terms of corrosion, brake-fluid leaks and dangerous, or potentially dangerous, faults. **ATTENTION!** *See my* **WARNING** *above, with the added provisos to keep clear of all moving parts and from the high-voltage ignition wires attached to sparking plugs.*

Your brake master cylinder is normally secured to the vehicle bulkhead inside the back of the engine compartment. *With your braking system still under the footbrake pressure,* you also need physically to double-check the security of its attachment to the vehicle frame. Use reasonable force to see if it will move up, down or sideways. It shouldn't. The surrounding bulkhead area through which your master cylinder is bolted must be of unyielding strength for at least 30cm all around, despite any minor rust.

Your master cylinder reservoir tide level must be above the minimum mark. This tide level can easily be seen through a transparent reservoir, but the trouble must be taken to remove the lid from an opaque reservoir, in order to check it properly.

Lastly, look for splits in the reservoir casing, or weeping of fluid anywhere in the vicinity. Note: a tide level a little below the maximum line does not necessarily mean there is a leak. Disc brakes have the characteristic of a falling master cylinder fluid level as the brake pads wear. This is allowed for in the max/min markings on the outside of the reservoir.

In **Step 2** I discussed in terms of lost motion the response at the road wheels to your turning the steering wheel. Even if you found no evidence of slackness earlier, you are now obliged to continue with under-bonnet inspection of your steering parts. If your vehicle has power-assisted steering, you need the engine running whilst you make these checks. **REMEMBER:** stay clear of hot or moving parts and high-voltage ignition wires.

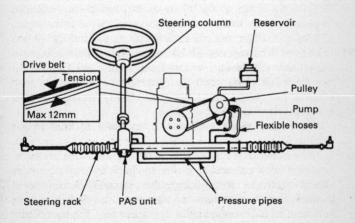

Fig. 3. Power-assisted steering.

Have your helper turn the steering wheel to and fro enough for the front wheels to begin to switch back and forth under load. While he does so, examine every link in the steering column visible from within your engine compartment. Some types go through more than one arranged "kink" before reaching the steering gearbox or the steering rack. Pay special attention to all pinch-bolt/U-clamp arrangements. Move on to look at security of mounting of your steering gearbox or rack. What you cannot see from above, you should be able to confirm from below, in **Step 7** shortly. A steering rack is shown in figs. 2 and 19. If you are not familiar with a steering gearbox, you will easily recognise one from fig. 20. On this, you need to check the top-plate bolts for tightness; they are prone to working loose.

You now turn to your power-assisted steering unit itself (if one is fitted) but *first switch off the engine.* Inspect the drive-belt. See fig. 3. The surfaces of the belt that touch the pulleys should not be *too polished or shiny;* otherwise the belt will tend to slip, especially at high speeds. There should be neither deep cuts nor fraying anywhere around the belt. Check the tension by pressing at the mid-point of the top run of the belt. The deflection of the belt when you do so ought to be about 12mm − neither too tight nor slack. Examine all parts of the system for excessive leakage, by which I mean fluid dripping. Because of the thinness of power-steering fluid, it has a tendency to seep past even tight joints, and therefore slight "wetness" around the reservoir, the pump, the hoses and pipes, and the unit itself is acceptable.

But fluid dripping must fail. That needs to be rectified immediately.

The pipework of power steering needs to be in good condition. There should be no flattening of the metal pipes, nor should there be any kinks. The flexible rubber hoses themselves should be leak- and cut-free. The power-steering reservoir must be topped up to the right level. Too much fluid induces leakage. Too little affects efficient operation. The latter is really a failure point; however, most testers would either top it up or advise the customer to do so urgently, rather than fail the vehicle.

Finally, examine the mounting of all parts of your power steering unit. Everything should be secure.

Continuing under your bonnet, look over your flitch panels. Where these under-bonnet side panels are load bearing, they must come under your most careful scrutiny in respect of corrosion. This is especially so when suspension units are integrally mounted and depending on such panels, or perhaps the attachment of a whole sub-frame assembly is involved. The mounting points, and such panels generally, must be secure and in no danger from rust. Refer to Chapter 4 to see how this may apply to your vehicle's construction; and look there for discussion of corrosion assessment generally.

If you have McPherson-strut front suspension, you must check the strut top mounting points on each side carefully, for security and for corrosion; however, a little bit of rust flake can be safely ignored. Recognition details and diagrams for this type are in Chapters 2 and 3. The big centre-nut at the top of

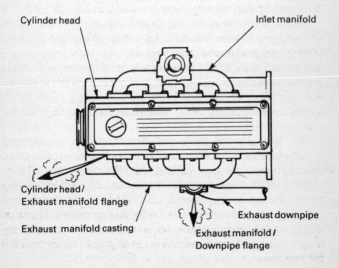

Cylinder head

Inlet manifold

Cylinder head/
Exhaust manifold flange

Exhaust downpipe

Exhaust manifold casting

Exhaust manifold /
Downpipe flange

Fig. 4. Exhaust leaks.

these, which is crucial to their locating properly, is prone to looseness and both sides must be equally carefully checked. If they have become only finger-tight, or move easily with very gentle persuasion from a spanner, then you need a torque wrench − a special spanner which can tighten to a precise degree − and a workshop manual. That manual will state the correct tightness, which you can only apply exactly by using this tool. Should that tightness not be easily achievable, it could be that the flexible rubber bonding supporting the top end of the strut has deteriorated to the point replacement or repair is required − a garage job. There are other causes leading to the same problem but they all require garage expertise to rectify. See figs. 24 and 29. These nuts are either of the self-locking type or of the castellated type. In the latter case they should be retained with a split pin.

Next, check that the engine has sufficient oil and water. Then ask your assistant to start the engine and let it tick over.

From hereon, again heed both my **WARNINGS** *given above in respect of hot, moving or high-voltage parts, etc.*

Listen straightaway for any exhaust blow (gas escape), all around the exhaust manifold close to the engine. See fig. 4. If you can get the palm of your hand close to where you think a leak is, you may be able to confirm a blow by feeling it. (This may be easier to do from below, during **Step 6.**) Another sign of a leak is a sooty deposit. In the early stages of a leak it may produce a useful clue − a 'phutt-phutting' sound at engine tick-over speed, made more obvious if the engine is revved up slightly for a moment. If your assistant briefly closes the tail pipe by holding a balled-up rag to it, this helps expose leaks. The engine should slow or stall when he does so.

BEWARE! *A manifold and exhaust system become burning hot within moments of starting the engine − so take care!*

A favourite leak-point is at the exhaust manifold/downpipe flange, where the exhaust pipe itself is attached to the manifold casting. Any leakage here tends to develop quickly into a major exhaust blow. Make certain the securing nuts/bolts are tight. The quiet 'burr' of your engine running soon becomes a loud, unpleasant racket. It is then illegal for you to drive the vehicle, and MoT failure is obligatory. Another commonly suspect place is at the exhaust manifold/cylinder head flange, and sometimes a manifold casting will itself crack.

There should be no leaks but just how much of a leak is needed before an MoT tester will fail it, is in the hands of the individual. His guidelines are to consider the safety of you and your passengers from inhaling fumes, and the normal "average" level of noise expected to be made by your vehicle as designed. So, if your vehicle is abnormally noisy or fumes could be getting inside, you need repairs. See **Part Two, Exhaust Repairs.** You may be able to undertake some repairs yourself.

Your last under-bonnet task now, is to check the fuel feed-line and the carburettor(s), or the fuel-injection system, for any dripping or leaks. You still need your engine running, so that fuel is flowing and leaks should show.

WARNING. *Do not smoke or use any naked flame. Petrol and its vapour ignite instantaneously.*

Fig. 4 shows an inlet manifold. The carburettor(s) would be attached to this, with the feed-line from the petrol tank at the back being easy enough to identify. Fuel-injector pipes feed into the engine cylinders directly adjacent to the inlet manifold. They can be traced back to their electronic control unit and, again, the feed from the petrol tank to there should be obvious enough. Whichever system you have, run your eye over everything slowly, remembering that evaporation of petrol quickly removes the evidence, and therefore any dark-stained areas deserve special attention. Since sparks (from the ignition, starter motor, etc.) can mix with petrol vapour with explosive results, your vehicle can be failed if the MoT tester believes there is any risk, even from a microscopic leak.

Step 6: You need now the front wheels of your vehicle raised off the ground for inspection underneath. The safest method is to use a pair of steel ramps available from motorists' D-I-Y stores. Access instead to an inspection pit, or to premises with a suitable built-in hydraulic vehicle hoist may be alternatives.

YOU ARE WARNED *that clambering underneath an*

insecurely jacked up car is potentially fatal. DO NOT EVEN CONTEMPLATE make-shift lifting schemes.

Note that the lightweight wheel-changing jacks provided by most small-vehicle manufacturers are really intended only for that purpose. The jacking points they utilise are prone to rust and failure as the vehicle ages, and the jacks themselves easily wobble over, especially if the ground underneath shifts.

So, never rely on such a jack alone. *Always back it up with solid blocks of wood, etc., so that, should the vehicle fall, it cannot pin you to the ground.*

For the moment, *all* your vehicle's wheels must be standing on a solid base, whether you are using ramps for the front wheels, an inspection pit or a professional hoist. Your handbrake must be on, and the rear wheels suitably chocked too.

From underneath, with the engine running, return to checking the exhaust system − from where you left off under your bonnet, through to the back tail pipe(s). Holes in the pipework or silencer(s) will not be passed; neither may suspect repairs be acceptable.

As you go along, look for secure attachment of the exhaust system. Waggle everything. (Old gloves may usefully protect you from heat.) An exhaust pipe will fail the MoT Test if its mountings − usually combinations of metal and tough rubber/canvas − are either torn, loose or missing, allowing the exhaust pipe to rattle. When satisfied all is well, you can stop the engine. D-I-Y repairs may be possible for some of the above exhaust faults I have mentioned. See **Part Two.**

Exhaust emissions are not easy for owners to check themselves. However, provided the engine is not in need of major overhaul, and burning litres of oil as a result (smoky carbons visible in the exhaust), the exhaust should be fine if the air cleaner is in good order and the engine is in good tune. If you are not an expert, armed with a workshop manual, tuning is best left to garages with their computer tuning equipment. Chapter 2 gives more information on exhaust emissions. The regulations do not apply to diesel-engines.

Step 7: Investigation of major steering components from underneath is your next target. Having your bonnet open lets light down to help you see but you need a powerful torch for this inspection as well. You need your engine ticking over if you have power-assisted steering.

If you identified lost motion in your steering during **Step 2** or **Step 5**, hopefully you will uncover the culprit now. Also remember to look again from below, at those steering items you may not have been able to see too well when looking down from above during **Step 5.**

Have your assistant turn the steering slowly from lock to lock *before* you go under the car, to make sure that when you do, there will be no danger of the vehicle slipping off its support. Once satisfied you can safely go underneath, ask him to rock the steering back and forth in quick succession, as well as turning it through its full lock-to-lock range of movement, whilst you then watch every inner and outer joint in your steering linkage from below.

Visible loss of communicated movement seen along the track rods, as they are called, or through any joint (track rod end), means your steering must be failed. On vehicles with a steering gearbox there are extra joints to be looking at, at the drop arm and at the idler arm. See fig. 20. Look at the security of attachment of the idler arm to the chassis or bodywork too.

Unacceptable wear anywhere will be most obvious as your assistant switches from turning one way to the other. It is important that he turns full lock to full lock, as sometimes faults only appear during a limited part of the turning range. Look for sloppiness due to vertical play in any joints too.

In this last respect take an extra close look at the steering swivels — top and bottom on some vehicles, bottom only on others — and, mainly on very old models, in the form of a single king pin. Look at the steering/suspension set-up pictures in Chapter 3, where there is also more information about what to look for. These top and bottom ball joints are usually made like a ball sitting in a cup. There must not be so much wear that, while the steering is being rocked from side to side, the ball lifts from its cup and then drops back into it. There should be no discernible lift (or drop if the arrangement is inverted). The ball should only have enough clearance to enable it to rotate in its socket. Note: some of these joints are even internally spring-loaded to keep them tight. In the case of king pins, there should be, if any at all, only a scarcely detectable movement between the king pin and the bushed (sleeved) parts of the stub-axle housing which rotate around it. (Chapter 2 and fig. 26 will help you see how a king pin/bush arrangement works.)

Lost motion can arise when the mounting of parts such as the steering gearbox or rack is insecure. So check that aspect as well − under the duress of being steered one way, then the other. Look for any serious corrosion or *fracture* of body structures to which such attachments are made. Remember, at least 30cm all around mounting points must be sound, free of splits, etc.

Retaining nuts or split-pins can go missing from steering joints − or be about to do so. They must be in place. Mostly these are only found on one side of a joint but on old vehicles they may be needed on both sides. Watch track rod ends, *all* swivel joints, locations of king pins, etc. (When a king pin is held round its *middle* as will be found with most beam-axles and single-arm suspensions, a locating cotter pin − at right angles to the king pin itself − must be present. This stops the king pin working loose and sliding out.) You must vigorously scrutinise the fundamental security of every joint in the linkage in this respect. More on this in Chapter 2 with figs. 19, 21 and 22.

You may discover basic stiffness in the whole steering, though this is difficult to judge, except in comparison with another vehicle of the same model known to be in good order. The problem could reflect partial seizure somewhere, power-assistance failure, or perhaps distortion or misalignment. The last two might be the result of an accident; for example, the underneath of the car may have hit a boulder whilst driving over uneven ground. Any bent or distorted parts must be replaced.

There should be no clonks or bangs during lock changes.

These are more easily detected if you hang on to a track rod while the steering is being turned – but watch your fingers don't get trapped! Nor should any parts (including your tyres) foul each other or the bodywork or chassis. Look especially for brake hoses being pushed aside by steering parts. (If it's happening, look for a maladjusted, loose or missing steering lockstop. That's the item which should prevent the wheels from being turned too far. Look at how much the brake hose may have become worn too. See Chapter 2.)

Whilst your assistant still turns your steering lock-to-lock, examine the gaiters on each side of your steering rack (assuming you have a rack and pinion system). Usually this is best done from underneath but you may find looking from above helps too. As the pleats of each gaiter open up, pull the rubber gently where necessary, to make sure you will find any tears or leakage of fluid coming from inside. A torn or holed gaiter must be renewed. (Otherwise the rack-lubricating oil drains away.) Both rack and pinion and gearbox-type steering arrangements eventually suffer from wear. See Chapter 2. On some cars, a steering damper may be fitted. These are explained in Chapter 2 as well, together with faults to look for.

Your steering is closely bound up with your suspension and, often, transmission driveshafts too. You should be aware that dangerous steering faults related to these other systems may not produce anything so easily noticeable as lost motion. However, I return you to these and other aspects in **Steps 9** and **10.**

Step 8: Stop the engine for the moment if it has been running. You now consider your brake parts and brake linings at all four wheels. For MoT Test purposes no dismantling is required. All the tester is expected to do about linings he has been unable to inspect is advise you of that fact, leaving you to make sure they are looked over later. (He should normally do this, for example, in respect of drum brakes – where linings are always hidden – even when the brake-performance test in **Step 12** is satisfactory.)

Fig. 1 shows a disc brake and a drum brake. If you have drum brakes you may prefer to look at the linings more thoroughly *now*, using your own knowledge, or with the help of a workshop manual. They need to be of minimum thickness, 1.5mm. For safety, once they're under 2mm, I would replace them. Linings which are bonded on to their brake shoes may be acceptable a little thinner than those which are riveted on because, on the latter, the lining face has to be 1.5mm *clear above the top of the rivets* rather than the platform of the shoe. A lining worn right down to the rivets, naturally, *must be renewed*. (On some designs the leading shoe − which does more work − starts life with a thicker lining than the trailing shoe. Check the specification before you rush to change trailing shoes "just because they look thinner". They may be quite all right.)

Disc brake pads, by contrast, are mostly visible, or can be seen with the aid of a small mirror. Technically, if the thickness of any pad has reduced below 3mm (or has reached the stage that a pad wear-indicator lamp is coming on) you must get it renewed or the tester ought to fail your vehicle. However, 3.8mm is the figure more usually worked to in the trade so, as a general guide, don't allow the thickness to fall below that. Should your disc brake pads be shrouded, you may, equally, like to take a proper look with the aid of a workshop manual, etc., and a small amount of dismantling at this stage.

Note: brake linings must always be matched for wear on either side of the vehicle. Renew linings at *both* front wheels or *both* back ones but never do so only at a single wheel.

At each wheel examine the disc or drum, as minutely as possible, for any sign of fracture, or contamination with oil or grease. (The latter aspect can have a disastrous effect on your brakes. A clue that it is happening may be tell-tale signs of oil splaying out from the centre hub of the wheel fixing, seen if you remove the wheel embellisher.) Also investigate the secure fixing of caliper housings (disc brakes) and/or brake backplates (drum brakes). A good shake should not reveal any movement in these items.

Step 9: You can now return to your brake hydraulic lines tracking each one onwards from where you last saw it under your bonnet. A powerful torch or a lead-lamp is essential. If your brakes are power-assisted, have the engine running — and remember my warnings about hot and/or moving parts, deadly exhaust gases etc. Fig. 1 shows where you can expect brake lines to run.

Your assistant must maintain very hard pressure on the footbrake pedal as you eyeball these hydraulic lines all the way to each road wheel. Especially include the entry unions (joints) where they reach the disc brake calipers or drum brake backplates.

In order to know the sort of faults to look for (beyond ones described earlier) *please refer once more to Chapter 2.*

A brake hydraulic system usually divides into appropriate directions at, or shortly after, leaving the brake master cylinder. Individual wheels may have separate pipes all the way from there, or they may share a pipe initially but one which splits its direction as it gets nearer to the front or back pair of wheels. There may well be dual pipework, or a partially duplicated system, for extra safety. The brake fluid reservoir is normally mounted integrally with the master cylinder as assumed in **Step 5**. However, it may be a separate item on a few older cars. On most designs, metal pipework runs along main bulkheads and panels until finally a flexible hose is used to reach each wheel. Where there is a rigid axle, flexible hose may be used to jump from the body to the axle instead of it having to be used for the final connections to the wheels.

A pressure-limiting valve is fitted in the hydraulic line to the back wheels in some vehicles. This serves to reduce the brake pressure on the rear wheels when necessary and minimise the risk of skids developing. If you find one, look for security of its mounting as well as for leak-free unions.

If your stop-light switch is installed somewhere in the hydraulic lines, as is common, look for leaks at each side, as well as at the valve itself which can also leak.

Next, take a methodical look around all the basic bodywork, chassis members and structure underneath your vehicle. Look for insecure, sharp-edged and/or jagged, or rusting-through, body panels or sills. Any item which is likely to be dangerous

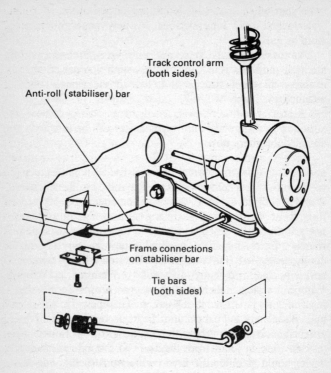

Track control arm
(both sides)

Anti-roll (stabiliser) bar

Frame connections
on stabiliser bar

Tie bars
(both sides)

Fig. 5. Anti-roll bar, track control arms, and tie bars.

will mean your vehicle must fail its Test. Examples might be metal that is torn or rusted through, or about to fall off or flap. How extensive the problem is, and whether other fixtures or danger to the public are involved, are the important criteria.

You are concerned with all load bearing areas. This may include floorpans if they form part of the inherent strength of the vehicle, or if a gaping hole presents a threat to safety.

Beware of rusty patches which may soon break through. These often start when leaks have led to the carpets above remaining waterlogged over long periods. A new panel might need welding in.

You are particularly concerned with box sections formed for strength purposes. On cars or vans with a separate chassis, examine all the points on it to which the vehicle body is attached.

Chapter 4 on corrosion assessment examines how the structure of differently built types of car can be affected. You should look there now.

You next turn to anti-roll (stabiliser) bars, track control arms and tie bars (the latter sometimes referred to as radius arms or radius rods). You are looking to see that they are neither bent, damaged, fractured nor severely corroded, and that their mountings are secure. Fig. 5 shows them. The secret here is not just to look — you could be deceived — but to shake these items hard. As with attachment points for all major components, the bodywork must be free of any weakness, severe rusting, or fracture, for at least a 30cm radius all round. With such crucial attachments I recommend you double that at least. Flexible metal/rubber bonding is often used at such joints and this cannot be passed if it is in a deteriorated or loose condition. Arm mounting pins must be neither bent nor loose.

Many vehicles have rear stabiliser bars to be looked at too. They should be obvious but you will also find them described under Panhard rod (one type) in Chapter 2, and shown by fig. 15. Examine them with the same attention to detail as above.

If you have a traditionally designed front-engined rear-wheel-drive vehicle, two important Indirect Test Items come next. See fig. 16. First you need to look for a perished rear mounting of the gearbox. If you can push the output housing upwards and then let go gently, a disintegrating rubber seat should at once be obvious. **WARNING:** *This can require heavy effort; only try if you are used to such work or you are able to effect some simple mechanical leverage,* but be careful you do not crack the box. Lift must be evenly applied — not just to one point. Second, you must physically check each universal joint (U.J.) on the propeller shaft. Grasp the shaft with both hands straddlng the U.J. and try to twist it one way, then the other. Any significant wear in a U.J. is easy to detect. Your grasp should straddle the U.J. to avoid confusing any slackness in the gearbox or the differential unit, with wear in the U.J. Two U.J.s are marked in fig. 16; sometimes there are more.

Returning to all vehicles again, you now ask your assistant to pull your handbrake (parking brake) on and off, while you look for free operation of the mechanical linkage. There should be no stiffness. First, however, make sure the wheels are still safely chocked so that your vehicle cannot move! Your engine must be switched off and, as an added precaution to hold the vehicle, an automatic gearbox should be set at PARK, an ordinary gearbox should be placed in 1st or reverse gear.

Frayed or badly chafed handbrake cables, bent or damaged rods or any sloppiness at the actuating levers in the system — especially where it reaches drum brake backplates or disc brake calipers — will mean MoT Test failure. Actuating levers are prone to being very sloppy.

Fig. 1 shows a typical handbrake open cable arrangement. Some work instead via sleeved cables as on bicycles, or combining the two types. You may find that yours operates on the front wheels using similar arrangements. Whatever design you have, the security of all cranks, levers, pulleys, etc., is vital. Retaining split pins must be intact. The load bearing structure on which each item is mounted must always be sound for at least a 30cm radius from the point of attachment to the extent the design allows this. The full range of movement required of any item must be available, so that there is no restriction on the operation of the handbrake. Where a cable passes through a sleeve which should be retained by a stop, that arrangement must be sound. The main pivot of the handbrake handle itself must not be about to wear through. It can usually be seen from underneath. On some vehicles — usually those having a dashboard-mounted handbrake — full inspection of the linkage will require looking under your bonnet too.

Move on to checking your fuel tank, and the fuel supply line, back up to where your under-bonnet inspection reached earlier, in **Step 5**. *Heed my* **WARNINGS,** *given there, about smoking and naked flame.* Look for loose fixings or chafing on the line. Have your engine running again (remember before you start it that you may have had it in gear) for the while, so that leaks which may only happen whilst the fuel is on the move, will show.

Next, concentrate on your suspension. At each wheel, look at the spring and its shock absorber − which reduces the "bounce" of that spring − and at the general way all the fixings relate to each other.

A few moments peering upwards from underneath should be enough to identify which arms, locating links, rods and so on, attach a wheel to the vehicle, and which ones steer or drive it, as well as to see how it is sprung − all without any need to know the names of the parts. Which parts have a dual suspension-and-steering role should also become clear. And, on some vehicles, you may notice how your driveshafts are also partly fulfilling these two functions. Most shock absorbers have a telescope-like appearance. They are sometimes mounted *inside* a coil spring. With what is known as McPherson-strut suspension, the shock absorber is integrated within the main strut (or leg) of the suspension instead of having a separate unit.

Chapter 2 has details of various common suspension systems, and a look there should help you understand how yours is constructed. Chapter 3 also has useful detail.

Focus your attention firstly on all *joints* in the suspension which need to swivel, or to move up and down. To confirm "where the action is" and discover problems, if any, have your assistant turn your steering to and fro (engine on if power-assisted), and, after, bounce your car up and down a bit.

You are looking for any bent, distorted or fractured arm, and for excessive wear in swivels and other joints. Some of these joints are bushed (sleeved); some may be constructed in bonded metal/rubber etc., designed to give a degree of flexibility. It should be obvious if such material is breaking down and loose, or perhaps, if a tightly fitting (though

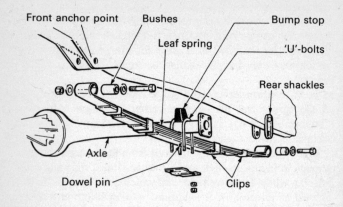

Front anchor point   Bushes   Bump stop

                        Leaf spring   'U'-bolts

                                        Rear shackles

              Axle

Dowel pin   Clips

Fig. 6. Leaf spring. Use of a lever detects wear.

movable) joint is suffering from some seizure.

Whatever the type, there should be no wiggling about within any of these joints at the front or the back wheels. Anything more than "scarcely detectable wear" must be failed. Locknuts, split pins, etc., must be in place.

You are also concerned with the strength of the vehicle structure wherever these movable suspension joints are attached to it. Clearly it's no use being hinged to some bodywork or chassis member which is about to collapse! Whilst the general principle of having a sound area of 30cm radius around any attachment applies, in the case of suspension you must use common sense and look round more widely. (For example — is the entire front or rear suspension mounted on a sub-frame which is itself fractured, distorted through a road accident, or about to fall off? — See Chapter 4. Note that the flexible bonded mountings of some sub-frames are prone to deterioration; they need rigorous inspection.)

You will return to additional scrutiny of some of these movable joints (with the vehicle jacked up to remove its weight, or the springing, from them) later. Meanwhile, examine your *springs* − your second focus of attention on the suspension at each wheel.

Springing at the front is most usually by coil. The same applies to the rear on many front-wheel-drive cars. However, leaf springing, in combination with a rigid rear axle, remains common for the majority of rear-wheel-drive configurations. Independently coil sprung rear suspensions are sometimes used. Then you will find "open" or exposed driveshafts connected to a centre-mounted differential. Rubber or fluid-based all round suspension systems are discussed in Chapter 2. Suspension by torsion bar is also explained. Should you have rigid front axle and leaf springs, treat them as for rear leaf suspension.

I go into leaf springs and axles in great detail below because, whereas independent coil springing means that each wheel is attached to the vehicle via arms and links of various sorts − all of which involve the joints you have been directed to look at in the last few paragraphs − a rigid axle (and therefore the wheels) is solely attached to the vehicle via the leaf springs. The security of the springs and their attachments, both to the axle and the vehicle, therefore assume even greater importance.

A coil spring, front or back, must be correctly located, and must not be broken or partially so. The seating above/below it must be sound. The spring must still be capable of compression (that is, the vehicle must not be sagging on its springs), and the bump-stop rubber must be in good order.

A leaf spring is shown in fig. 6. Look for a cracked leaf or one that has come out of line with the others due to broken clips, etc. Examine for secure fixing of the spring to the axle, as well as for any problem of it slipping along, or round, the axle. The U-bolts must be secure and tight, and the split pins, locknuts or whatever, which retain them, must be intact, doing their job properly. See also Chapter 2 if you find problems. Consider whether you have a sagged spring which allows the vehicle almost to rest on the bump stop. (You may need to look at another vehicle of the same model to compare.)

Turn now to the (leaf) spring mountings. At the front

mounting (known as the anchor point) is a bush, consisting of a hard rubber block encircled by the spring, and having a metal sleeve at its centre through which the attachment bolt to the bodywork passes. At the rear mounting are shackles incorporating two pairs of softer rubber bushes. These are not usually sleeved. One pair is encircled by the spring, the other is mounted on the bodywork. Each pair has an attachment for forward and aft movement of the spring in use. As with all suspension mountings, look for sound chassis or bodywork members covering an extra wide area around these attachment points.

Look to see if the sleeve within a front bush is worn or has developed an elongated or enlarged diameter, thereby creating a large clearance between it and the securing bolt inside. Use a lever to prise the eye of the spring (the part of the spring in which the bush is fitted) away from the adjacent bulkhead. This will reveal undue wear. Inspect rear shackle bushes in much the same way. Check that the spring eyes themselves are not split or broken.

A small amount of play (looseness) in a bush is acceptable. However, you cannot have a bolt "rattling about", or ignore the danger that it could soon shear (break off).

You are also interested in bush wear if it has led to *sideways* play of the spring. That is, if the spring eye has become able to shift along the shank of the bolt in either direction, so that the entire spring could begin to skew out of line with the car. Again, appropriate leverage applied will check this out. Any substantial movement being possible would have to be failed. Note that loose U-bolts holding the spring to the axle can place exceptional strain on the front and rear mountings, leading to the sort of faults just described. Tap these U-bolts with a hammer to see if they are loose.

The third and last focus for attention on your suspension from underneath at each wheel will be your *shock absorbers*. I described these earlier. More information about different types is in Chapter 2. Look for excessive fluid leakage, physical damage, severe corrosion or loose fixings. Try to twist or shake the shock absorber in order to detect faulty mountings. A flexible dust cover can conceal fluid loss. Press around it to see if copious fluid oozes out; it should not. Slight fluid loss, damage or corrosion may not have damaged the *functioning* of a shock absorber; you will be testing that next. If that is OK your vehicle should pass.

Step 10: Return your vehicle to the ground. (If you have had the luxury of a pit to work from so far, you won't need to!)

To test the shock absorber action at the four corners of your vehicle, bump each wing up and down in turn; do so about four or five times with ordinary adult strength. Always release the wing *quickly*, and at the *end* of a downward stroke. In every case, the wing should return to its normal resting height in one or one-and-a-half movements. No further see-sawing is permitted. That would show that the shock absorber was either weak or it had failed. The fault can make a vehicle very difficult to control, or even lead to skidding. Fig. 7 shows this bump test.

You now need to jack up each wheel in turn so that you can:

1) Look at tyre valves, treads and *inner* walls (those facing the centre of the vehicle).

2) Test wheel bearings.

3) Check steering swivels again and security of stub-axles.

4) Look for slackness in driveshafts (where applicable).

5) Look again at brake hoses, shock absorbers, suspension springs, underneath the (wheel) arches, etc., taking advantage of angles of view which may not have been possible so far.

WARNINGS! Jacking up a vehicle can be *dangerous*. Refer back to **Step 6**, page 37, for safety precautions. The jacking up process gives you a unique opportunity to observe all joints in the basic assemblies to which each wheel is attached

Fig. 7. Shock absorber bump test.

(including the attachment points of those assemblies to the chassis or bodywork). If you watch them as the vehicle rises you may well notice play or slackness which cannot be seen at any other time.

Where you jack up front wheels is critical so far as steering swivels are concerned. Look now at Chapter 3 to identify your type of suspension and see exactly how it must be jacked up.

Taking either of your front wheels first — jacked up safely and in accordance with Chapter 3 — go through the five points above:

1) Inspect the valve carefully. Then have your assistant rotate the wheel slowly while you examine the tread and the inner wall. Because of stringent tyre laws, most people are familiar with tyre faults that make them dangerous or illegal. However, details are provided in Chapter 2 should you wish to double-check what to look for.

2) To test the wheel bearing, start by spinning the wheel and listening for any sound of rough running. (A front-wheel-drive car will need to be in neutral. A front-wheel-handbraked car will need the handbrake off. **N.B.** Are the other wheels chocked securely?)

There should be no obvious grating or rumbling noises. Then grip the wheel at six o'clock and twelve o'clock and rock it firmly, towards and away from you. **WARNING:** *Both tester and assistant must be careful because the vehicle wheel is jacked up. Wooden blocks or substantial supports should be in place so that, even if the jack fails, human parts cannot be trapped.* You should detect a tiny amount of "end play" (or end float") between the wheel hub (to which the wheel fits) and what is called the stub-axle (to which the wheel hub is fitted). However, this must remain in the "scarcely detectable" category. Anything more will be failed.

If there is excessive play, you will want to establish if it is in the wheel bearing or in the steering swivel joints. Ask your assistant to apply the footbrake. If the play disappears with the brake on, then the trouble is likely to be in the wheel bearings. However, if end float persists despite the brake being on, then the wear is probably at the steering swivel joints or a king pin/bush is worn, unless it is coming from loose wheel nuts or some other hub fault.

3) Wear in the steering swivels (or ball joints), or on old vehicles in the king pin or its thrust washers, may become more obvious if you ask your assistant to rock the wheel again as in 2) above, while you take a hawk-eyed examination of these parts. However, further examination is still needed: *Firstly,* using a heavy lever under the tyre as in fig. 8, lift

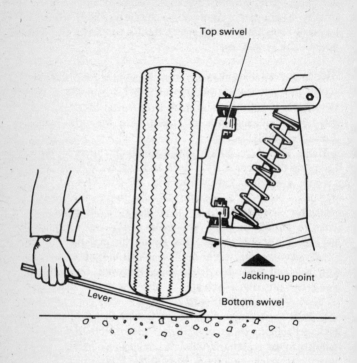

Fig. 8. Levering wheel to check for wear in joints.

and let go. Each top and/or bottom joint should lift as one, without gaps appearing between ball and cup. In a king pin arrangement you should be able to detect undue wear on the king pin itself or in the thrust washers that usually support and position the stub-axle on it. (What a king pin looks like, how to judge a worn thrust washer, and so on, is explained in Chapter 2, along with further information on more modern swivels, ball joints, etc. See also figs 22 and 26.) Anything more than scarcely detectable separation in any type will normally be failed, and the fault will have to be put right, usually a garage job. *Secondly,* watch the swivels while your assistant turns the steering gently full lock to full lock. Again, noteworthy problems should show.

If you have discovered from Chapter 3 that you have McPherson-strut front suspension, slightly different tests are required to reveal wear of this sort. Refer to Chapter 2; please refer there if you have this type.

4) On front-wheel-drive vehicles, the inboard flexible universal couplings of the transmission driveshafts may deteriorate due to internal looseness, fractured U-bolt clamps, or general wear − or even, perhaps through oil contamination or splitting, perish altogether. The shaft itself may become distorted or bent. Additional descriptions of these joints are in Chapter 2. Check which of the two main types you have.

The universal knuckle joint into the road wheel (also called the constant velocity joint or C.V.J.) may also start to run slack, or its rubber gaiter or "boot" can become split or detached, allowing grease out and dirt in. Any of the above faults are due cause for an MoT Test failure.

With the gearbox in neutral (and a front-wheel handbrake released if necessary − in which case rear wheels must be chocked!), have your assistant turn the road wheel slowly a whole turn several times, while you inspect the gaiter, the shaft and the inboard coupling for the above problems.

Next ask him to turn the road wheel back and forth gently. At the point of switching over from forwards to backwards you should most easily spot looseness in the joints. Let him make the switch-over point at at least four equally spaced different places in relation to one whole turn of the wheel; that way you should see the joints from every possible angle. To help you see the constant velocity joint gaiters properly, have the wheel first on one full lock, then the other; this will enable you to fan out the pleating with your fingers and make sure there is no serious deterioration, splitting, misfitting etc.

Now place the vehicle in gear and have him switch the wheel back and forth (i.e. fore and aft) the little it will go. So that your engine CANNOT start *make sure your ignition is off!* This will allow you to establish if there is significant *lost motion* between the wheel turning, and that turning causing the gearbox/engine to try to turn. There is always some lost motion, and it may be hard to assess if yours is normal, without having seen another matching vehicle known to be in good order. However, it should be sufficiently obvious if any joint is suspect enough to fail its MoT Test — or if it is even about to collapse — which is always possible! You can also carry out an additional test on your C.V.J.s later, on the ground on the move, if you have the slightest concern. See Chapter 2.

5) Ask your assistant to press hard on the footbrake (the engine must be running if your vehicle has servo-assisted brakes) while you take this, additional, detailed look at your brake hose — the flexible part arriving at the wheel — and its unions. There should be no weeping fluid; the structure of the hose should be undamaged. Chapter 2 describes the sort of faults to expect. Look there to be sure.

If you haven't been able to before, it is worth giving the shock absorber a shake to see that its fixings are tight. Double-check for weeping fluid on it too.

Lastly, cast your eye around the whole wheel arch area for anything else which may be amiss, for example, a broken or unseated suspension spring or a perished bump stop.

Now lower the vehicle to the ground before moving to the other front wheel. Jack that one up safely — again, in the right place according to Chapter 3, and, as before, watching all joints carefully during the jacking up process.

Then repeat points 1-5 above on that front wheel.

Move on to the first rear wheel. You will be looking at:

1)   Tyre and valve.

2)   Wheel bearing.

3)   "Open" driveshafts and differential, if applicable.

4)   Brake hose, shock absorber, suspension spring, etc.

Jack the wheel off the ground safely. See my **WARNINGS** at **Step 6**, page 37. Again, watch joints as the vehicle goes up.

1)   Inspect the valve and the tyre exactly as for the front wheels, especially the normally out of sight inner wall.

2)   Spin the wheel while you listen for a rumbling bearing. You need the handbrake off unless it happens to work on the front wheels. You need the other wheels securely chocked and the gearbox in neutral. There should be no untoward noises.
Rock the wheel as for a front one. There should be just detectable play, if any at all, on a back wheel. Second, grasp it at 3 o'clock and 9 o'clock and attempt to push it away from you (towards the opposite back wheel) and back again. On vehicles with a rigid back axle this is to see that the wheel bearing is not sliding along on the half shaft. On vehicles with "open" driveshafts from a centre differential, this test may reveal any slackness in the shaft links (universal joints, etc.). A microscopic amount of end float detected this way will be OK. But anything easily measurable would have to fail.

3)   Some rear-wheel-drive vehicles have, as I have mentioned above, "open" driveshafts from a centre differential, instead of having a rigid back axle. As with front-wheel-drive driveshafts, these have universal joint couplings to be checked for *lost motion* and/or disintegration! So, in addition to the partial test above, follow the same method of looking over these driveshafts as in 4) for the front wheels — except that you won't need to turn the steering!
The secure mounting of the differential unit itself needs looking at too. Very slight movement of a differential casing may be allowed for in the design but if you find any alarming degree of movement, you will need to check this out in a workshop manual or with the vehicle manufacturer or his

agent. Obviously perished flexible mountings would be cause for MoT Test failure.

4)   Follow the routine given in 5) above for front wheels.

Now lower that rear wheel to the ground before moving on to the last wheel. Jack that one up safely as before, watching joints as before, and repeat points 1-4 as above. Finally, lower the vehicle to the ground again.

Before a road test in **Step 11**, stand away from your vehicle. Observe whether, with the steering straight ahead, any of your road wheels appear to be leaning — that is, in or out beyond the normal characteristics for your model. (You may first need to drive the vehicle briefly to allow the suspension to settle after coming off the jack.) If you have any concern, see **Stub-axle assemblies** in Chapter 2. Also consider whether the cause may be a sagging or misaligned spring, or a damaged swinging arm, etc., more especially in respect of rear suspension. Remember that accident damage may have distorted a major suspension part, or the area on which it is hung.

Step 11, Road Test: Since the introduction of roller brake testers, a road test has become largely optional in the MoT Test. However, they are still made at testers' discretion, as well as by small, remote garages who are exempted from having to have the special equipment, and they have to be done on those vehicles which, perhaps because of unusual design factors, cannot be machine tested.

So many faults can be discovered or confirmed during a road test that I, personally, believe they should become compulsory again. **I feel sure that a majority of mechanics testing vehicles on a daily basis would agree.**

I advise *you* to take a test drive and be thinking about what follows here whilst you do so.

It is extremely difficult to pin-point all faults during a static MoT Test. For example, it may not be possible to check an inaccessible upper shock absorber mounting. However, that mounting is likely to clonk loudly when driven on the road. Similarly, a front or back wheel bearing, which yielded no real clue as to anything wrong when the wheel was spun or rocked, may drone when in motion, indicating that seizure or collapse is not long off. Here are some more faults which may emerge, though they are not intended in any way as a complete list:

1) Steering pull: Assuming the tyres are all right, incorrect steering castor angles, or excess toe in or toe out of the front wheels (according to the design), can badly affect handling. Instead of tending towards straight running, as it should, the car may continuously try to veer away. Umpteen problems only show up when the steering is steered along a road. They may be rooted in a damaged front sub-frame, in the suspension, in the brakes, in a stub-axle or elsewhere in the steering. Whatever the source, the tester must consider how relevant the fault is to the MoT Test, and decide accordingly.

2) Steering wander: This describes when a vehicle no longer responds in the normal precise relationship to the steering being given by the driver. It may be due to excessive play in the front wheel bearings. On vehicles with front disc brakes, this may be accompanied by an increase in brake pedal travel. Other causes might be soft front or back tyres, or incorrect steering geometry (see Chapter 2), or poor suspension.

3) Steering wobble: When the steering wheel vibrates in your hands at higher speeds, this may arise from unbalanced wheels or badly seated tyres on their rims.

4) Tyre "blister": Large "blisters" or bulges in the tread make a noise to listen for, peculiar to themselves. Imagine the situation when a chunk of tyre tread (say, 15cm long by 12cm wide) parts company from the inner bracing of the tyre — somewhat in the raised fashion a human skin blister develops. As the wheel rotates, this loose area will hit the road with a distinct slap at regular intervals. The sound is that of a soft

hand clap. When there are bulges on the tread areas of *both* rear tyres, the vehicle may rock slightly from side to side as the bulges encounter the road surface. This is more pronounced at very low speeds and may disappear altogether at high speeds. It should not be confused with a steering wobble which is felt specifically at the steering wheel.

5) Vehicle vibration: This can be caused by unbalanced front or rear road wheels. However, in a rear-wheel-drive car it more usually derives from a propeller shaft being out of balance. Either the shaft itself is out of balance or the universal joints on it are worn or loose. An exhaust rattle can be mistaken for vibration too.

6) Drive faults: Loose transmission parts can lead to clonks during the take up of the drive, or, nasty noises when your accelerator is released on the run. Sometimes the reason is merely an excessively high engine tick-over. But however serious the cause it must be found. An old gearbox may jump out of gear under load.

7) Brake faults − see **Step 12** below.

Step 12: Brake efficiency I come to last because private owners are not generally equipped to measure it. You may already be aware that your brakes have been falling below par or that they are becoming overdue for adjustment; or earlier **Steps** here may have alerted you that all is not well. Otherwise, an "emergency" footbrake test (conducted with due care bearing other road users in mind), during your road test, should be enough to confirm whether your brakes are up to the mark. A test from 30-40 m.p.h. on a good dry surface should suffice.

In an MoT Test the tester will first ensure that the tyres are at the correct pressures in order not to give misleading figures on his brake roller test machine. (On the road *you* must have the pressures right to prevent avoidable skidding.)

He will also consider (as must you) the general state and age of the vehicle. He may feel justified, after his underbody corrosion inspections, to fear sub-frame collapse, or perished brake pipes bursting. (As the vehicle is "driven" on the roller machine and the brakes are applied, this is a real possibility on a rusty old vehicle, because full brake application, necessary for the Test, puts extra stress on these items.) If he does, he may refuse to carry out the brake test. He will then fail the vehicle on the grounds that he has been unable to complete your MoT Test, and note his reason on the failure form.

If you fear that your car or its brakes will collapse next time they are needed in a hurry, and you are still driving it or considering testing the brakes on the road, you don't need an MoT Test, you need a psychiatrist! Need I say more?

However, when you carry out a brake test on the road in the ordinary way, a swift, smooth, progressive stop should result. Uncharacteristic (for the model) skidding, steering-pull or veering to either side, or snatch or judder of any sort, or any other oddity, mean you need to seek expert advice.

As to handbrake efficiency, technically your handbrake should alone be able to hold the vehicle on a minimum slope of 16% at least (that is 1 in 6.25). For a rough check, apply the handbrake fully with the vehicle on level ground. Either you alone or two of you should try to push the vehicle forward. It shouldn't move. If it moves, see to the handbrake.

2

A-Z OF TEST ITEMS IN DEPTH

This chapter complements the **12 Steps** of Chapter 1 by expanding background knowledge and detail. Main subject headings are in A-Z order; the index below should help you quickly to find the page you require.

Brakes

ABS and other anti-skid systems

ABS stands for Anti-Lock Braking system. ABS, from Bosch of Germany, was the first major anti-skid braking system. There are now several types, including Lucas Girling's SCS − or Stop Control System. The efficient functioning of these anti-skid devices is not a concern of the MoT Test

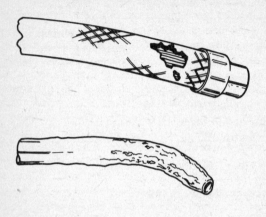

Fig. 9.   Brake hydraulic line deterioration.

because the brakes still work conventionally if the anti-skid arrangements fail. However, the warning lights and controls for these systems are required to be in working order.

Brake hydraulic lines

The amount of brake fluid in a braking system is pre-determined in order to transport a given amount of force when called upon to do so. The system's efficiency depends on the full amount of fluid being available. The loss of even a tablespoonful or two can be critical.

Poor workmanship assembling brake pipework often leads to problems, though the trouble may not reveal itself for many thousands of miles. This can even originate at the factory. But it is more likely to be due to unskilled repairs, perhaps following accident damage. I have even seen dangerous re-routing of pipework causing it to be fouled by moving parts when the steering was turning, or whenever the vehicle encountered bumpy terrain.

Cracks in the outer cover of brake flexible hoses are enough to justify MoT Test failure. See fig. 9. These hoses support a lot of pressure. You cannot just wait until they give up the ghost. To test them, pull back any guard (such as a coil spring over a brake hose, which protects it from stone chips), and bend them (the hoses) over, to expose any cracks. The ends are especially susceptible to cracks, so you must take a critical look at each of them, as well as in between. You are concerned with any *swelling under pressure* and about rotting or fraying affecting the material. Hose which has been contaminated by oil, grease etc., can go soft and even squidgy. Feel the whole length of these pipes thoroughly. Any suspect length must be renewed. Hose with its inner core showing will fail. Guards must always be repositioned correctly afterwards.

All metal brake pipework must be securely clamped to the vehicle underframing, and the rear axle (when applicable), respectively. Brake pipes hanging loosely are lethal. Imagine a low-slung brake pipe getting caught by an object in the street; the predictable outcome of sudden loss of braking could be very unsightly. But *any* loose fixings are a danger. They eventually lead to fractures, kinks, splits, etc.

Hydraulic leaks occur mostly at brake pipe unions (see fig. 1). Every one should be checked thoroughly. At these unions, and at the joints where the pipes reach a drum brake backplate or a disc brake housing, threads may have been stripped, or overtightening can have spoiled the way delicately shaped parts mate together. Twisted pipes often develop hairline cracks. Joints which have succumbed to exterior rusting frequently suffer undue force as an operator tries to free them off. Add to these snags the fact that any job done on the hydraulic system must be followed by bleeding − a specialist process to eliminate any air which has entered the brake fluid − and you can see why I do not recommend the D-I-Y driver to try to put right hydraulic leaks himself! One mistake on a brake repair job can be lethal. So, if you find the slightest leak or weeping unions, go to a reliable garage and

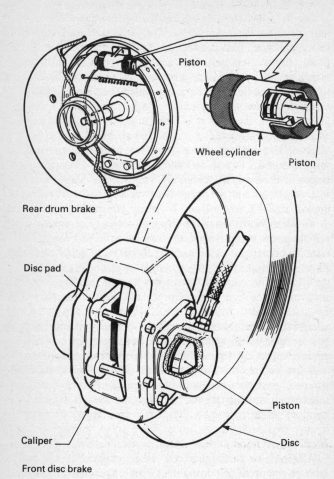

Piston

Wheel cylinder

Piston

Rear drum brake

Disc pad

Piston

Caliper

Disc

Front disc brake

Fig. 10. Wheel cylinder, drum brake. Disc brake piston.

have them repaired.

A very common fault is brake pipes corroding at the point where the pipes pass through the flitch panels (under-bonnet side panels) to the front wheels. They should be well-supported on either side of such panels and a rubber grommet should protect them at these sites, but these often perish or go missing. A vehicle will fail the Test if a perished or missing grommet, or insecure mounting, causes, or is likely to cause, a brake pipe to chafe the vehicle frame.

Brake pipe corrosion

Second to bodywork-corrosion assessment as the greyest area of the MoT Test is determining corrosion on a brake pipe. Superficial corrosion should not be penalised. However, it is vital to establish properly whether corrosion is superficial or deep-seated. To test a brake pipe, it should be scraped thoroughly down to bare metal with a penknife or similar tool, although this is not necessary if the pipe looks new or fairly new. If the penknife quickly reaches smooth, shiny metal, the pipe is passable. However, if the surface is badly scored and pitted with rust and indentation down into the thickness of the metal, then the pipe is not serviceable. After sound pipework has been scraped for a test area, a film of ordinary car grease should be applied to the bare metal to preserve it from further rust attack. (In an ideal world all brake pipework would be protected thus or some better way in the first place.)

Some testers are casual about brake pipework and do not scrape away rust. Owners, who do not want to be victims of slack work which could lead to accidents, should inspect their own pipework thoroughly for their own peace of mind.

Brake roller tester

A seized, or partially seized, wheel cylinder is a common fault on motor vehicle hydraulic drum brakes. Rear ones seem to be more prone to seizure than front ones. The wheel cylinder(s) contain(s) a piston, which is a sliding part designed to transfer the pressure or force you put into the hydraulic line (with your foot when you press the pedal) directly onto the brake shoe. There are equivalent piston arrangements within a disc brake. See fig. 10.

A brake roller test or a road brake test should reveal such faults. With the garage brake roller test, the dials will register significantly low readings at wheels where there are problems. On a road brake test, the vehicle will not respond quickly to the effort exerted on the pedal, and it will tend to pull to one side if only one wheel is affected — especially (but not only) if this is at the front of the vehicle.

An ABS or other anti-skid system may require all four wheels to be rotating at the same time for the device to be tested properly; however, the underlying braking system can still be tested on a single-axle brake roller tester, and, as explained earlier, this is all that the MoT Test requires.

Whereas braking efficiency can be expressed as a percentage directly from the readings on a brake roller tester, on an MoT *road* test a decelerometer is used instead to measure the percentile braking force. They are explained in the next section.

More often than not, when a vehicle owner is issued with a failure certificate, it records the vehicle's braking efforts. For example, handbrake effort might be 24% and footbrake effort 55%.

The law is only concerned to see that a safe minimum percentage effort is present. The pass requirements are: for a footbrake 50%, for a handbrake 25%. However, only 16% is required of the handbrake if the hydraulic braking system on the vehicle is of a dual-circuit type.

Referring back to my example failure certificate above, the footbrake passed but the handbrake failed because the hydraulics were not dual-circuit.

A brake roller test demands applying the footbrake and, separately, the handbrake, as hard as possible. This action may reveal faults an owner may be unaware exist. A wheel cylinder which was beginning to seize up before, unnoticed because the owner never needed to apply the brakes very hard, may show up or become completely seized during a brake roller test. A handbrake cable which appeared perfect previously may snap. A corroded brake pipe may burst; a badly cracked brake hose may spring a leak. The same faults could emerge during a road brake test. Therefore it is important for owners to appreciate the fact that such problems occur surprisingly frequently during an MoT Test; and that they should *not* instantly blame the tester!

Decelerometer

Before the advent of brake roller testers, decelerometers were the chief means used to measure braking force. A decelerometer measures braking effort (the rate at which the vehicle slows down) and expresses this into a percentage which the tester reads off its gauge. The meter, though small and portable, is heavy enough to stay still when placed on the floor of the vehicle for the purpose of the road brake test. It locks on the maximum reading it registers, to enable the driver to read it afterwards. If the tester wishes to repeat the test he simply resets the gauge to zero first.

The test is undertaken from not more than 20 mph. The footbrake must be applied smoothly and progressively, not snatched, jabbed or slammed on, otherwise the reading will be inaccurate. The handbrake is tested similarly. The required percentage efforts are the same as given four paragraphs above.

Brake servo

Most modern vehicles have servo-assisted brakes. Contrary to popular belief, a servo unit does *not* make a brake more powerful. All it does is to help the driver apply the brakes; it increases only the force he exerts on the brake pedal. But this makes it easy for him to achieve maximum braking with little effort. Any force exerted by the driver on the brake pedal is multiplied at a given ratio — usually one and a half times.

Though, in theory, you would expect a failed servo unit to render the brakes inefficient and dangerous, in practice some vehicles can still be braked reasonably easily without it. Others cannot. Nevertheless, a vehicle with a failed servo must fail the MoT Test.

To test your servo, see **Step 2** in Chapter 1. For minor

repairs, see **Part Two.**

Handbrake

When excessive on/off lever travel is noticed on a handbrake which operates on drum-type rear brakes, it will usually be the brakes that need adjusting, not the handbrake cable. Where a handbrake operates on a disc-type brake, such excess travel will usually be slack in the cable, or its pivots, or at the handle fulcrum itself. Slackness of a handbrake connected to self-adjusting drum brakes will arise if that self-adjusting mechanism is faulty. When it is not the brakes themselves that need adjusting to correct slackness in a handbrake cable, remember that these cables must only be adjusted when absolutely necessary; otherwise unnecessarily stretched cables may be the only, and unwished for, result. When a handbrake cable is equipped with grease nipples it should be greased occasionally or as recommended by the manufacturer.

Exhaust emissions

When petrol is burnt by the engine, one of the gases that comes out from the exhaust pipe is carbon monoxide. This gas raises environmental interest because it is poisonous.

The aim of the regulations on exhaust emissions is to reduce the carbon monoxide (CO) content of the exhaust gases without unduly affecting the performance of the engine. If the exhaust is black and sooty, then too much CO is being emitted. If the emission looks grey or colourless, then less CO is being emitted and the chances are it will be all right, especially if you cannot see it. Thus you can usually see straightaway whether your exhaust emission is likely to be unacceptable.

The regulations also require the hydrocarbon (HC) content of the exhaust gas to be checked. Hydrocarbon here means unburnt fuel which has found its way out through the exhaust pipe. This HC test is not so much a pollution control one as it is to help ensure that engines are running properly when their CO content is checked. As mentioned in **Step 6** of Chapter 1, you are therefore well-advised to be sure your engine is in good tune when you present the vehicle for its MoT Test.

During the MoT Test, an exhaust gas analyser will be plugged into the exhaust pipe of your vehicle with the engine running. This machine will display the required readings to be noted by the tester. The analyser test applies to all petrol-engined cars, light goods vehicles and other vehicles first registered before 1st August 1975 falling within the main MoT testing scheme, although certain engine designs, such as the Wankel rotary one, are exempted. The exempted engines and those of the old vehicles built and registered before 1st August 1975 must be inspected instead to see that they do not produce "excessive" (black) smoke from their exhausts. This visible "excessive" smoke test is based on the premise that an exhaust gas with an excessive CO content both looks sooty and seems large in volume. The MoT tester will use his skill and experience to make a judgment in each case.

The analyser tests become more stringent for younger vehicles and, in due course, those fitted with catalytic converters will come under the strictest rules of all.

Exhaust systems

Poisonous exhaust fumes entering a vehicle affect the driver's alertness to a lethal extent surprisingly quickly. A leaking exhaust which has become excessively noisy may breach noise emission regulations as well. The presence of either problem will lead to MoT Test failure.

Front-wheel-drive transmission shafts

The inboard (engine end) universal joints (U.J.s) on a front-wheel-drive transmission shaft are commonly of two types — one made up mostly of rubber, the other made up mainly of metal (see fig. 11). These joints don't go on forever, although that can be the impression on better-built vehicles.

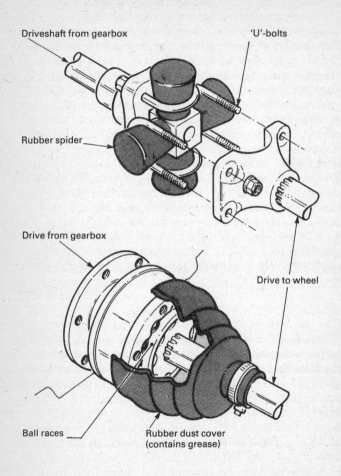

Fig. 11. Front-wheel-driveshaft inboard joints.

To test a rubber-type coupling, insert a lever into the coupling so that it can be "opened up" a little. If there are cracks in the rubber spider, a new coupling must be promptly fitted. Note: when fitting these couplings, the wheels of the vehicle must be resting firmly on the ground before the *final* tightening is done. This causes the couplings to assume a proper working position once tight. Failing this, newly fitted couplings will split within a short space of time.

The metal type are popularly known as "pot joints". Though generally more durable than rubber couplings, these joints can, after considerable mileage, become sloppy and need replacing. To test one, you need the vehicle up on ramps. Hold the driveshaft, next to the joint, firmly with one hand while you grasp the pot joint with the other. Try to rotate the joint first backwards (towards the rear of the vehicle) and then forwards (towards the front). If there is a lot of movement in the pot joint in this fore-and-aft rocking, then the joint is unserviceable and should be renewed. Ideally, there should be no more than just detectable movement — that is, less than 3mm "rock-on-the-pot". Anything more is unacceptable. On the road a really sloppy pot joint will give a dull clonk as the drive is taken up.

At the road-wheel end of each driveshaft is what is called a constant velocity joint (C.V.J. for short), as shown in fig. 12. This joint allows the front wheel to be steered and driven simultaneously. A rubber gaiter protects the joint. It also harbours an abundance of grease necessary for lubrication of the joint during service. If there is any split or tear in this gaiter, which looks much like a soft pouch, it must be renewed immediately. Replacing the gaiter is a garage job unless you are expert. If the gaiter is only torn, it *can* be replaced on its own; however, since dirt will probably have found its way in quite quickly, you will be well advised to have a careful internal inspection of the joint itself before deciding whether that should be put back, or replaced too.

An excellent test of a constant velocity joint for serviceability is for the vehicle to be driven around in circles on full lock, in each direction in turn. A clonking noise emanating from the front end during this test indicates a worn-

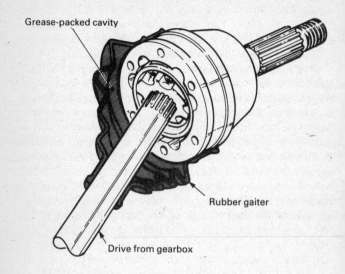

Grease-packed cavity

Rubber gaiter

Drive from gearbox

Fig. 12.   Front-wheel-driveshaft constant velocity joint (C.V.J.).

out C.V.J. It may be difficult to pin-point which of the two is
faulty, because it is sometimes difficult to place where a
"clonk-on-lock" is coming from, and, of course, both could be
faulty and noisy at the same time. Normally, however, a clonk
means a worn joint on the side to which the steering is being
turned. You often have to trim your hearing carefully to
establish the truth for certain. You may need an assistant to
drive the vehicle slowly round on full lock while you walk
alongside listening.

At the end of the splines that secure a driveshaft to the

constant velocity joint is a groove and in it a circlip or spring ring. This circlip stops the driveshaft parting company with the constant velocity joint especially on *full lock;* mechanics must always ensure this circlip is correctly reinstated after a repair job.

Headlamps

There are three basic types of headlamp design. The first is the European type, which may be round, rectangular or trapezoidal in shape, as shown in fig. 13. Their hallmark is that they have moulded into their glass face a stepped or "staircase" pattern. Their focus is checked on dipped beam.

Both the other categories are known as British American, but one group is checked for aim on high beam; the other group is checked on dipped beam.

British American types which are *checked on high beam* are round, and are marked either 1, 1a or not marked at all. These are shown above each other on the lefthand side of fig. 14.

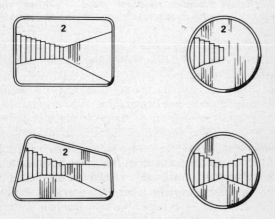

Fig. 13. European headlamp styles – all checked on dipped beam.

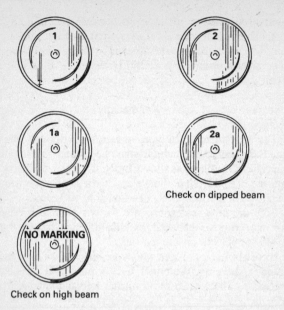

Check on dipped beam

Check on high beam

Fig. 14. British American headlamp styles – lefthand column checked on high beam; righthand column checked on dipped beam.

British American types, which are *checked on dipped beam* are marked either 2 or 2a, as shown above each other on the righthand side of fig. 14.

Note: when, occasionally, one headlamp of a twin-headlamp paired set-up has only one beam, it has to be checked on that beam.

Headlamp aim (focus)

MoT testers use a beam setter to check your headlamps' aim. You should be able to make an adequate test for yourself, and adjustments when necessary, as explained in **Part Two.**

Insecure body panels

When, in a tester's considered opinion, a body panel is

either so corroded or so insecure *as to be likely to fly off* at any time, causing injury or damage to other road users, the vehicle must be failed. This is the only reason for any body panel, *other than a load bearing structure,* to be failed. A panel which *is* also a load bearing structure and which is severely corroded should, however, fail even though it may still be relatively secure. This includes panels onto which other essential items are mounted – e.g. a stabiliser bar, a Panhard rod, a leaf spring or a seat-belt mounting. Such bodywork cannot be passed if its security is in doubt.

Panhard rod

A rear leaf spring acts both as a suspension unit and to

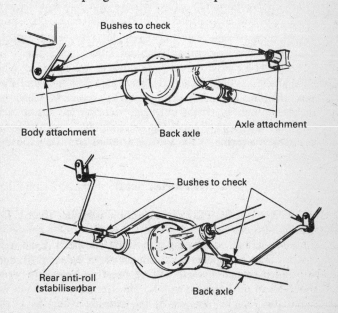

Fig. 15. Panhard rod, and rear anti-roll (stabiliser) bar.

prevent sideways movement of a rigid rear axle. When coil springing is used in conjunction with a rigid rear axle, it cannot prevent this sideways movement. Therefore any sideways movement has to be controlled by fitting some form of stabiliser such as a Panhard rod.

The difference between an ordinary stabiliser bar (also known as anti-roll bar) and a Panhard rod is that the main function of the former is to minimise the vehicle "rolling" from side-to-side, whereas the main function of a Panhard rod is to prevent the rear axle moving sideways; however, they both perform each other's function to a limited extent and manufacturers may choose between various set-ups. Fig. 15 shows an example of a rear anti-roll (stabiliser) bar and of a Panhard rod.

A Panhard rod can either be mounted transversely and to the rear of the axle as shown in fig. 15, or, two bars can be mounted longitudinally (that is, in line with the vehicle length), one in each corner, travelling from the axle to the vehicle frame. The body mounting points for either type must be of sound metal for at least 30cm all round, unless designed to be fitted to a lesser but specially designed area (in which case that must be entirely sound). From a tester's point of view, the bushes through which the bolts fix, securing the rod at each end, must be in good condition. An axle that moves side-to-side makes steering of the vehicle difficult and dangerous.

Propeller shaft

In Chapter 1, **Step 9,** I discussed how propeller shaft U.J.s. (see fig. 16) can be tested. Worn U.J.s can throw a propeller shaft out of balance, causing it to vibrate at speed. Even with unworn U.J.s a propeller shaft can get out of balance. An out-of-balance propeller shaft will be most noticeable between about 50 m.p.h. and 60 m.p.h. The tremor will be felt all round the vehicle, not just at the steering wheel as might happen with a steering vibration caused by a front wheel wobble. An out-of-balance propeller shaft can be corrected by

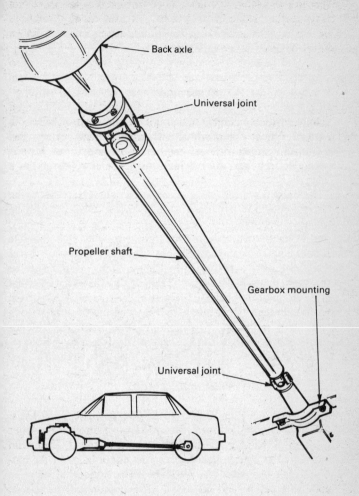

Fig. 16. Propeller shaft and universal joints.

balancing the propeller shaft just as road wheels are balanced. This is a specialist job, not a D-I-Y job. It is noteworthy that badly out-of-balance rear wheels can also cause the whole vehicle to vibrate; however, they would have to be very badly out of balance before this happens.

Shock absorbers

The correct technical name for a shock absorber is a damper. When a road wheel encounters a bump, it sets the suspension spring into an oscillating motion. That is, the spring deflects past its original position and rebounds in a

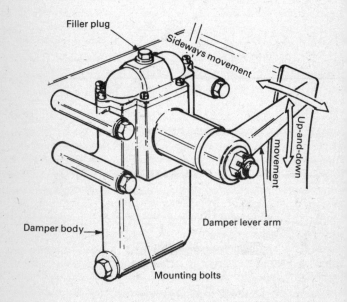

Fig. 17. Lever type shock absorber.

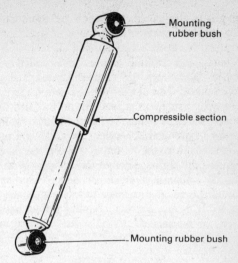

Mounting rubber bush

Compressible section

Mounting rubber bush

Fig. 18. Telescopic shock absorber.

continuing series of bounces because of the energy given to it as the result of the bump. You experience this as a bumpy ride. The purpose of a damper is both to minimise such discomfort and to help avoid skidding, which could be triggered were the up-and-down motions to continue unchecked. The damper reduces the spring oscillation to the barest minimum by absorbing the energy to the spring.

A damper is fitted to the vehicle in such a way that it is set into action every time its road wheel encounters a bump. One end attaches to the vehicle frame, and the other end to a suspension or axle part that moves vertically as the wheel strikes a bump.

There are two basic types of damper − the lever type and the telescopic type. See figs. 17 and 18. Both are filled with a thin, mineral-base oil. Both dissipate the energy they have absorbed by pumping this oil from one chamber to another through fine orifices. This action creates resistance and therefore controls the deflection of the road spring. If the damper is designed such that only the rebound action of the spring is controlled (that is when the compressed coil spring is re-opening, or the deflection of a leaf spring is on the return), it is called a single-acting damper. If it controls both the bump and rebound actions, it is called a double-acting damper. Sometimes the damper is designed to resist both strokes but to control the rebound action more.

Excessive sideways movement on the arm of a lever type damper (see fig. 17) must be failed. When new, there is no such sideways movement. Therefore any sideways movement must be failed unless the tester is certain the slackness is not hindering the performance of the damper. With the telescopic type, the mountings must not budge when the damper is grasped and an attempt made to rotate it. If one of them does, it probably means that that mounting rubber has weakened and it must be replaced.

Steering

Description of types

There are two sorts of steering arrangement on modern vehicles. These are the rack and pinion systems and the steering gearbox types. Figs. 2 and 19 show rack and pinon set-ups, and fig. 20 shows a steering gearbox assembly. All early vehicles had steering gearboxes of various kinds. Nowadays, the rack and pinion alternative, which is cheap to produce and gives the driver a more direct "feel", has largely taken over for lighter vehicles. But it is not suited to heavy vehicles, so they continue to be fitted with steering gearboxes.

New steering parts

You should be aware that even brand new spare parts can sometimes be faulty! I have, on several occasions, found that brand new track rod ends and steering swivels were unserviceable. I then had the situation of convincing those customers, because the problem was so unexpected.

Power-assisted steering

Power-steering fluid is pumped under pressure to operate

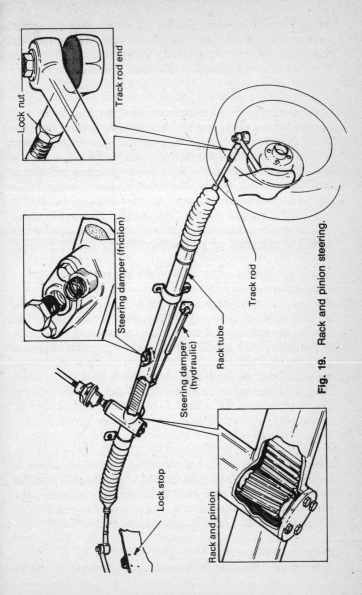

Fig. 19. Rack and pinion steering.

Track rod end

Lock nut

Steering damper (friction)

Steering damper (hydraulic)

Track rod

Rack tube

Lock stop

Rack and pinion

the assisting mechanism. Should the blood-red or transparent light green fluid ever find its way into the steering rack and pinion gaiters — where it would mix with the rack-lubricating (engine) oil — the system would need immediate repairs.

Rack and pinion steering system

A rack and pinion system enables the turning of your steering column to push a serrated, or gear-toothed, shaft (the *rack*) from left to right, or vice versa, inside a tube. A *track rod* joins each end of the rack to its respective wheel hub. As the rack moves, each wheel is pushed or pulled accordingly, so that the two front wheels are steered in unison. In fig. 19 you can see how the pinion (gear) at the bottom of the steering column meshes with the rack. You can also see how two rubber gaiters, one each side, are fitted to the ends of the steering rack tube to retain (engine) oil lubricant for the rack and pinion. Track rod ends carry ball joints. (The inboard ends are hidden within the gaiters.) These allow for the necessary universal angular movements which you can visualise must be necessary, both as your road wheels are turned for steering and when they have to be able to move up and down with the suspension. They are further discussed in the rest of this **Steering** section.

A steering rack wears out after a very high mileage (say, anything over 60,000 miles, but it can wear out sooner). Wear develops between the pinion and the rack teeth (see fig. 19). Also, wear can occur between the rack shaft and the bearings that support it in its tubing. A truly worn rack will rattle, particularly on rough roads. The noise is similar to that of a worn shock absorber.

Steering damper

Steering dampers are not fitted on all vehicles; however, they are part of the rack and pinion steering system on many vehicles and earn MoT Test failure if unserviceable.

The purpose of a steering damper is to absorb the vibrations to which the steering is subjected, so that they either don't reach the driver or are minimised before they do. It is simply a telescopic shock absorber placed transversely along the

steering rack. See fig. 19. One end of the damper is secured
to the rack and pinion arrangement and the other end is fixed
to a point on the vehicle frame such that the damper forms a
narrow "V" with the rack. As the steering is turned from left
to right and vice versa, the damper opens and closes, just as
do vertically mounted dampers on suspensions.

Steering dampers can leak or eventually wear out. Look for
serious oil leakage from a damper seal, for which it would have
to be failed. To test its efficiency, disconnect one end of it from
either the rack or the vehicle frame and open and close to feel
the resistance. It should be roughly equal to that of a bicycle
hand-pump when the tyre is almost full. If not, it is worn and
should be replaced. It will also be failed if its body or fixings
are badly damaged or corroded to the extent that breakdown is
obviously imminent. In a simpler format, inset in fig. 19, a
damper consists of a sprung pad(s) − or tensioner − acting
directly on or under the rack. These rarely give trouble.

Steering gearbox

A steering gearbox enables the turning force exerted by the
driver at the steering wheel to be transmitted through 90° to
the road wheels. It is also a leverage system. The gears allow
the driver's effort to be multiplied by approximately 10 times
(10:1) for light vehicles, and by roughly 30 times (30:1) for
heavy vehicles. Usually there are more turns required to move
the steering wheel from lock to lock as compared with a rack
and pinion system − another reason why those systems are
generally preferred for lighter vehicles.

Whereas there is only one possible gear arrangement in the
rack and pinion steering design, a steering gearbox has five
different individual gear types to choose from, all in common
use. These are: worm and sector, screw and nut, recirculating
ball, cam and peg, and worm and roller. It is outside the scope
of this book to go into detailed descriptions of these; suffice to
say that different manufacturers prefer different designs.

Fig. 20 shows a steering gearbox and steering parts. The top
of the box has a removable bung through which it is filled. It
is important that the box is full at all times. Consult a service

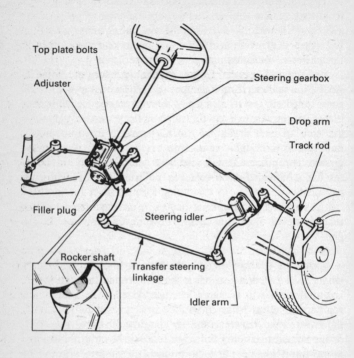

Fig. 20.   Steering gearbox and steering parts.

or workshop manual to see what the correct lubricant is, if you find it needs replenishment.

Some designs have provision for shims (packing) under the top plate, used as a means of adjustment for wear which may develop after long usage. Others have an adjustment screw instead. However, steering box adjustment is a sensitive job, and, unless you know what to do, it must be done by a reputable garage.

A vertical shaft is found inside all types. It is called the rocker shaft. This extends out through the base of the gearbox and connects directly to a drop arm − again, found on all the five types. The drop arm connects to the transverse steering linkages and track rods.

A steering idler arm will also be found towards the opposite side of the vehicle from the steering gearbox. However, as the name suggests, it has no turning force function; its vital task is to *carry* the ends of the transverse linkage and the track rod that side, in such a way that they can move/turn appropriately in unison as the vehicle is steered. All the joints in the steering gearbox and idler arrangement need your careful scrutiny as discussed in **Step 7**, Chapter 1, and additionally in the next few pages below.

The steering gearbox does not require any maintenance apart from ensuring there is sufficient oil in it at all times. The top plate bolts must be secure and tight. After a very high mileage − or if it has been allowed to run dry for any appreciable time − wear develops in the gearbox. Such wear can often be seen when, with the road wheels on the ground, the steering wheel is turned gently to and fro. The rocker shaft, instead of just rotating either way smoothly in its bearings, rocks sideways at the same time. This slackness may be adjusted out using shims or the adjustment screw mentioned above but it is an expert's job. If full adjustment does not remove the sloppiness, then the steering gearbox must be renewed or overhauled. Bearings and mating parts within the gearbox simply becoming worn out from use can also necessitate renewal or overhaul.

Steering angles or geometry

To compensate for the forces that act on the front wheels of a vehicle when it is in motion, the front wheels are, looking from the top, either tilted in or out, *slightly*. They are also tilted either forward or rearward *slightly*. The small angles of tilt concerned are called the camber and castor angles respectively. Collectively, they are known as the steering *geometry*. In addition, the front wheels' direction is either *slightly* angled inwards (toe-in) or angled outwards (toe-out). This latter aspect is known as the *wheel alignment*. The combined effects of the geometry and the alignment make certain that, in motion, the wheels run true (roll in a straight line), and this prevents tyre wear which would otherwise be ruinous to tyre and pocket alike.

The settings can be put out gradually through steering components getting worn. They can also easily be distorted by impact — for example, by the driver hitting kerbs hard or thumping down into large potholes, and in accidents. Roadholding becomes less positive as a result. The vehicle no longer responds as directly as it should to the steering — a condition called steering wander. Tyre wear accelerates rapidly.

It is up to the vehicle owner to ensure his front wheels are correctly aligned and that the steering angles are not distorted (perhaps through damage during previous ownership). Unfortunately, in the nature of the MoT Test, it is unlikely for an MoT tester to detect such faults unless they are pronounced. A road test of your own may help pinpoint them. Keeping a careful watch on your tyres for signs of uneven wear can save you a lot of money by giving you early warning of trouble.

You need a trustworthy and properly equipped garage to have the steering geometry and wheel alignment checked (and corrected if necessary). Because there may be no immediately visible signs that they *have* been checked or adjusted, it is easy for mechanics to skimp the work, or for swindlers to pretend it has been done when it hasn't. The first the hapless owner may ever know is when rapid tyre wear has occurred, but by then it is too late to lay the blame on a rotten garage!

Steering column couplings

We are here concerned with *flexible couplings, pinch bolts,* and *universal joints*. See fig. 2.

Flexible couplings tend to be ignored or not to be given the necessary inspection they deserve, simply because they are not always accessible. The steering should be turned from lock to lock and every surface of any rubber or leather coupling thoroughly inspected for cracks developing into the material, or for weakness. Such faults require MoT Test failure. The parts concerned must be replaced promptly. Testers dare not compromise on steering components.

Pinch bolts are, I'm afraid, frequently overlooked by mechanics during servicing. Therefore it is of paramount importance for testers to check them. A badly loose pinch bolt can cause the two halves of the steering shaft that it has jointed together to slip relative to each other or even disconnect, with catastrophic results. Always ensure that a pinch bolt is tight. To test for looseness of a pinch bolt and/or for chewed splines, see **Step 2** of Chapter 1.

Universal joints in the steering column serve a similar purpose as the U.J.s on propeller shafts. (See also fig. 16.) In this case, they allow the steering turning force of the driver to pass through an angle on its way to the steering rack or gearbox. Rust around these joints is the giveaway sign of wear. **Step 2** of Chapter 1 deals with testing them.

Steering ball joints and trunnions

Steering levers or rods are linked together by ball joints which allow universal (or angular) movements. If steering levers were joined rigidly they would bend or the joint would break as the road wheels encountered different contours of the road. When these ball joints are fitted to the steering linkage, they are called *track rod ends*. When slightly modified and fitted on the front stub-axles to enable each front hub to swing as the steering wheel is turned, they are called *top* ball joints if fitted at the top, or *bottom* ball joints if fitted at the bottom.

Fig. 21 shows how a track rod end is made up, and how a top or bottom ball joint goes together. A ball (also known as pin) sits tightly but movably in a bearing socket (or "cup"), and the bottom of that cup (even if it is in an upside-down position) is tensioned with a spring. On earlier versions, a grease nipple is fitted so that the pin can be greased, and an adjustment screw is available to adjust out any slackness as wear occurs between the pin and the bearing socket, but modern types are self-lubricating and non-adjustable. If the pin becomes slack in its

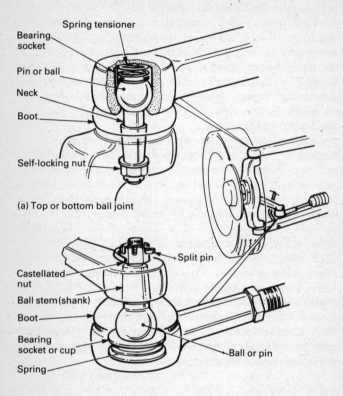

Spring tensioner

Bearing
socket

Pin or ball

Neck

Boot

Self-locking nut

(a) Top or bottom ball joint

Split pin

Castellated
nut

Ball stem (shank)

Boot

Bearing
socket or cup

Spring

Ball or pin

(b) Steering track rod end

Fig. 21. Steering ball joints.

socket on these, the ball joint has to be renewed. However, with the modern materials used, they do last a long time.

A rough description of a steering trunnion is a "threaded ball joint". The top and bottom ends of each stub-axle mounting/housing (the whole of which turns with its front wheel as you steer) are extended into shafts and their ends are threaded. To accept these threaded equivalents of a ball/pin, the corresponding "cups" or trunnions (one attached to and forming part of the upper suspension arm at its outside end, and the other similarly attached to the lower suspension arm) are internally threaded. The trunnions themselves (although fixed to the suspension arms) are not screwed fully tight or the steering cannot turn! As a rule, they are screwed right up and then backed off 1-1½ complete turns. The amount by which they are backed off (given in manufacturers' workshop manuals) is what enables the stub-axle housing to swivel in its trunnions. The bottom diagram of fig. 22 shows how a bottom trunnion is screwed upwards onto a link pin (threads covered by a rubber boot) and then connected by a fulcrum pin to the lower suspension/wishbone arm. The link pin rises through the whole arrangement and is similarly held at the top either to the shock absorber arm or to a secondary suspension arm.

Another trunnion arrangement can be seen in fig. 25. It is essential that, however trunnions are located to the suspension, all associated nuts, bolts and pins are obviously present and correct. If you have any doubt, comparison with another vehicle of the same type, or a look in the manufacturer's workshop manual should resolve them.

Steering king pins

In the steering arrangements just described above, the complete wheel/stub axle assembly is held betwixt two points – the top and bottom ball joints or trunnions. To allow you to steer, the whole lot is able to swivel or rotate at those joints. In a king pin arrangement a fixed *king pin* is incorporated into the outside end of a single suspension arm. That pin is so arranged as to pass through upper and lower bushed (sleeved) tubes, which are an integral part of the stub-axle casting. The

entire stub-axle assembly and front wheel therefore —
although tightly held by those bushes — can pivot round on the
king pin as you steer, in much the same way as a hinge pivots
round its centre-pin.

Fig. 26 shows simple king pin steering and how the cotter
pin mentioned in **Step 7**, locks the king pin in place. Here, the
stub-axle connection is to a beam-axle type suspension (non-
independent suspension) found on heavy vans and trucks. The
assembly itself is (as is usual) supported *vertically* on the king
pin by what are known as thrust washers. The nut securing the
narrow end of the cotter pin must be in place.

In **Step 10** of Chapter 1, I discuss how to examine king pins.
If, during that testing, you find that the stub-axle casting (and
all carried with it) is able to slide significantly up and down
the king pin, then it is likely that thrust washers must be worn.
However, because slight wear on thrust washers alone is not
critical to steering performance, they are best left and only
replaced when the entire king pin/bush assembly comes to have
to be dismantled because of wear elsewhere. (A good MoT
tester should not fail a worn thrust washer if all else is OK
unless that wear is extremely bad.)

Wear "elsewhere" on a king pin/bush means either the
bushes have worn thin and/or the king pin itself has become
stepped (indented). Such wear, discovered during the **Step 10**
testing already mentioned, showing anything more than just
detectable play should be failed. However, in general, slightly
more play is allowed on king pin/bushes than on ball joints,
and you should find that MoT testers use sensible discretion
when there is very little wrong.

Steering linkages security

Since steering linkages are crucial (i.e. an accident will be
extremely likely if they come apart), they are connected in
such a way that that ought never to happen. This is normally
achieved by the use of *castellated* nuts or *self-locking* nuts.

A castellated nut is shown in fig. 21, lower picture. A hole
is drilled towards the outer end of the thread across the stem
(shank) part of the ball, to receive a split pin. This nut is slotted

across its flats (hence the name castellated) to receive the pin. Here is how the connection is done: first, the track rod end ball stem or ball pin (without the nut) is pushed into the hole in the steering arm (coming from the wheel hub) and the castellated nut installed; then, the nut is tightened until one slot across the flats, lines up exactly with the hole in the thread; lastly, a split pin is pushed through the hole and its openable ends are then splayed out, thus preventing the nut from coming adrift. Your inspections need to establish that all such split pins are secure, with both their open ends still intact.

At times a self-locking nut is used in place of a castellated nut. This is a nut with a thin integral plastic sleeve thread-insert (having a slightly smaller diameter than the thread itself) towards what will be the outer end of its thread. As the nut is tightened onto the ball stem (shank), this sleeve grips the ball stem (shank) tightly and so prevents the nut from coming adrift. It can be undone with a spanner but will not loosen on its own. These nuts should not be re-used (new ball joints always come with fresh nuts). If you were to find evidence of a self-locking nut having run loose on its thread it must be replaced.

A track rod end is attached to the track rod with a locknut, as also shown in fig. 19. There is no normal reason for these to come loose unless the joint has been disturbed, but it is worth looking them over.

Finally, on the subject of security, an excessively worn ball may have developed a weak, reduced in size, neck. If the neck has become indented or worn down to any unusual degree, the track rod end will need replacement. Compare with others of the same model if in doubt.

Stop lamps, sidelamps, indicators

Stop lamps, sidelamps and indicators should be checked for correct operation frequently. The matter cannot just be left until the MoT Test! Best done by enlisting the help of an assistant, this can also be done by "sighting" them while you

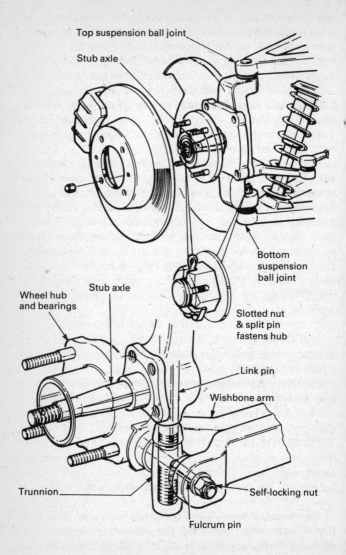

Fig. 22. Stub-axle assemblies.

wait in a traffic queue, for example, in "mirrored" shop windows or reflecting off the vehicle behind. Many vehicles have dashboard warning lights to alert the driver of a failure, but it's wise not to trust them completely.

Stub-axle assemblies

A stub-axle is the structure on which the front brake and the road wheel are attached (see fig. 22). It must be examined for bending, probably caused by an impact. Look out for leaning wheels beyond the normal characteristics for your vehicle. The cause of a leaning wheel may be, at first sight, inexplicable. However, further examination may well reveal a distorted stub-axle. Other causes could be corroded and/or buckled suspension, or badly worn steering top and/or bottom swivels. Some possible symptoms of a damaged stub-axle are abnormal and/or excessive tyre wear and the vehicle proving difficult to steer.

Suspensions

Leaf-spring, coil-spring, torsion-bar and *McPherson-strut* suspensions are about the commonest types fitted. *Hydrolastic, hydragas* and *rubber* suspensions have gradually been falling from favour. Nevertheless, all are discussed during the next few pages.

Coil-spring suspension

Though the McPherson-strut suspension also uses a coil spring, the suspension dealt with here is that which normally incorporates "coil-spring" in its name. Fig. 8 shows the type. Essentially, the coil spring is placed between a lower and an upper arm, on which are carried the steering swivels.

A coil spring does sometimes break in service. This is the first thing the tester looks for. Top and bottom abutment areas of a coil spring are so designed that the spring locates into special recesses to stop it rotating during service. A tester

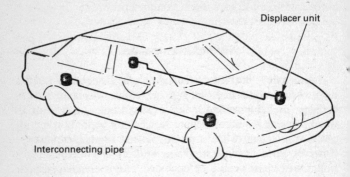

Fig. 23. Hydrolastic/Hydragas displacer layout.

should ensure this positioning has continued correctly, though the spring is unlikely to shift unless it has been wrongly assembled. Coil springs also sag. Once a spring is so weak that the vehicle becomes conspicuously lop-sided or abnormally low, the vehicle deserves to fail its MoT Test. It is a major repair which a garage should undertake.

Hydrolastic and hydragas suspensions

These combine fluid and rubber (hydrolastic), or fluid and gas (hydragas), to form a suspension. Hydrolastic or hydragas displacer units are mounted in each corner of the vehicle, and interconnected by pipes. Most often one pipe links the two lefthand side units together, the other links the righthand pair. Figs. 23 and 27 show the layout. When a front wheel encounters a bump, fluid is displaced along the connecting pipe to the corresponding rear unit, and vice versa when a rear wheel encounters a bump. Each unit has a built-in damper valve to control the flow of fluid leaving it. This control acts as a damper, replacing the need for an ordinary shock absorber. The displaced fluid reaching the rear unit acts on an internal diaphragm and raises the rear of the vehicle to the level of the front wheel. When the front wheel drops back to a level position, the fluid returns to the front unit and the vehicle settles to its previous riding position. Other types act slightly differently and only the rear displacers are linked, in this case together *across* the vehicle; the front ones each act independently.

Whatever type you have, you must examine the full lengths, as far as possible, of the interconnecting pipes for security of fixing, corrosion and leakage, and the displacer units for solid mounting and for rattles. Faulty displacer units rattle when the vehicle is on the move, especially when driven over bumpy roads. A rattle is indicative of internal damage of the displacer unit, necessitating replacement. The units are not repairable and are therefore replaced as a whole; this is a garage job.

In service, even without any leak in the system, the height of the vehicle off the ground can go down. To restore the proper height requires the system to be pumped up with the proper fluid and then bled of any unwanted air. Dealers for the makes concerned are all equipped to do this. If doing it fails to raise the height of the vehicle to normal, the unit is faulty and must be replaced.

Leaf-spring suspension

Before inspecting the leaves of this type of spring, road muck or mud must be scraped away so that each leaf and all the fixings can be examined properly.

When leaf springs become very weak they can reverse their shape from concave to convex, even though they may not always suffer broken leaves in the process. The vehicle owner is strongly advised to renew them, preferably in axle set, at this stage. (Most testers, myself included, do not like to fail leaf springs until there are broken leaves. However, I always strongly advise customers to renew very weak springs and I would feel obliged to fail a vehicle which had insufficient remaining clearance at a spring bump stop.)

U-bolts are used to clamp the leaf spring to the tubular axle.

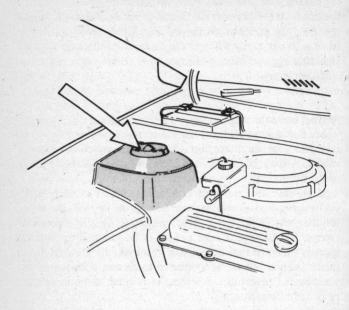

Fig. 24. McPherson-strut top mounting.

The bottom section of the axle is machined flat at that point to mate well with the spring. For additional security, this flat section of the axle is drilled, but not to any great depth, to receive a dowel pin which then goes through the centre of the leaf spring in order to keep the two components absolutely together (see fig. 6).

The U-bolts must be inspected for tightness and, as far as is reasonably practicable, a tester must ascertain whether the dowel pin (above) is intact or broken. This he should do during his brake roller test. With the rear wheels on the spinning rollers and the footbrake applied abruptly, any movement by the axle either sideways or along the length of the spring will show that the pin is broken. You can make your own check this way. Have your assistant try to push the vehicle forward while the hand- or footbrake is kept applied firmly. Watch the spring and the axle as described above. (What the assistant is doing here is trying to push the spring away from the axle while it is held in one position by the brake.)

Front leaf-springs are checked in the same way as the rear ones.

McPherson-strut suspension

See figs. 24 and 29 for recognition of a McPherson-strut and its component parts. Popular because of its relative cheapness of production, the McPherson-strut suspension is also simple and versatile.

One major fault — which only occurs after high mileages (unless due to faulty manufacture) — can be that the shock absorber part of the system develops a leak. Either the seal that retains the damper oil in the tube (see fig. 29) becomes unserviceable and springs a leak; or the highly polished piston inside the unit becomes scuffed through dirt getting past the seal, with this, in turn, leading to leak. Most manufacturers protect the highly polished piston with a rubber gaiter as shown in fig. 29. Although a torn gaiter would not be failed, customers are usually advised to get it renewed as soon as possible. This is a garage rather than a D-I-Y job.

A second major fault, again which doesn't happen until after very high mileages, is for the highly polished piston to get so worn that it rocks or is loose-fitting within its cylinder.

To detect this fault, jack up the vehicle anywhere solid on the underframe but *not* under the track control arm, and grasp the wheel at the 12 o'clock and 6 o'clock positions. Rock the wheel hard and watch for any movement at the point where the piston enters its cylinder casing. There should be no sideways movement of the piston in the cylinder. Any movement here (though you may consider a barely detectable one negligible and ignore it) means the strut must be renewed. Otherwise handling of vehicle will be affected and it must fail its MoT

Test. (**WARNING!** Readers are reminded that most ordinary tool-kit, wheel-changing jacks are not intended to be sufficient support for working underneath a vehicle. Always support the vehicle additionally with solid wood blocks, or similar items which cannot crush or topple, before going underneath to work or even before rocking a jacked-up wheel.)

Another, but unusual, fault is that the coil spring may become insecure in its seat (top or bottom). Turn the jacked-up wheel slowly from lock to lock and watch. If the coil spring shifts about or rocks while the wheel is moved from side to side, it must be put right.

The top of a McPherson-strut is supported by a bearing on which it swings. It is usual to replace or service this bearing at any time a McPherson-strut is removed for major repairs. Should this bearing become partially seized, the vehicle would have a tendency to pull to one side. To test if this is the case, find a wide, safe, clear road, and drive in a straight line at about 50 m.p.h. (80 k.p.h.). Loosen your grip on the steering wheel briefly. If the vehicle pulls markedly to one side (more than you might expect as a result of any road camber), a seizure may be the cause. However, unequal tyre pressures (front or back), steering geometry out of true, or other suspension faults could also be the cause, just as could be brakes binding and/or a wheel bearing(s) dragging. A very slight steering pull may be judged reasonable but any vicious pull must be failed. Rectification of this fault is a garage job.

The bearing top mounting can become loose. This was dealt with in **Step 5** of Chapter 1. Fig. 24 shows the mounting from above.

The McPherson-strut suspension employs only one steering swivel joint at each side of the vehicle. This is situated at the bottom of the strut concerned. See fig. 29. The top end of each strut is attached to the respective wing valance via a bearing as mentioned above. Because a McPherson-strut set-up integrates steering and suspension more fully than most other arrangements, it can be difficult to detect wear in these steering swivel ball joints at the base of each strut. There are therefore four special tests to make on them:

1) While the weight of the vehicle is on the ground (or safely on a ramp), have your assistant rock the steering wheel gently to and fro just enough to start the road wheels turning. If your vehicle has power-assisted steering, your engine needs to be running while he does it. You watch the joint to see that the "ball" part of it in no way separates from the "cup" part. (See fig. 21). The ball rotates in its socket but there must be no discernible separation going on.

2) Still with the wheel resting firmly on the ground (or ramp), lever the ball joint up or down (depending on design) with a suitable heavy implement. Do so by levering the track control arm (see fig. 29) onto which the ball or cup is attached, applying your leverage a few centimetres away from the joint. Lever a little and let go. Place your lever in such a way as to try to separate the "ball" from its "cup". Again, no appreciable separation of the ball from its socket should manifest itself, otherwise garage repair will be required.

This ball joint may be fixed either way up. Which way is unimportant. Your aim is to discover whether the ball can in any way shift from its socket. You will need to pivot your lever against the inside of the wheel rim, or against a suitably positioned solid object.

3) You now need to park your vehicle, with the road wheel on the side being checked parallel to, and touching, a high kerb. Then have your assistant turn the wheel so that the tyre becomes pressed reasonably hard against that kerb. Having done so, he must let it go *suddenly* whilst you are watching the swivel joint. The tyre should spring back a fraction and, as it does, you should be able to spot any substantial wear in the ball joint. Repeat the process two or three times in order to be quite sure there really is no sloppiness there. (**WARNING!** Watch out for passing traffic; do this only on a very quiet road.)

4) Now jack up each front wheel in turn, taking care to position your jack correctly each time (see fig. 29, Chapter 3). At each side ask your assistant to grasp the wheel at the

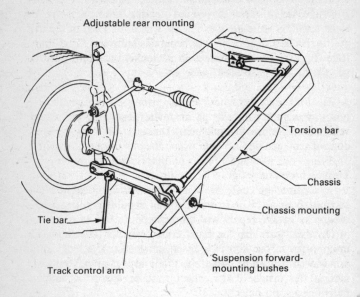

Adjustable rear mounting

Torsion bar

Chassis

Chassis mounting

Tie bar

Suspension forward-mounting bushes

Track control arm

Fig. 25. Torsion-bar suspension.

6 o'clock and 12 o'clock positions and rock it hard in and out.
Observe the joint while he does so. Again, any of the same
slackness I have discussed above will lead to MoT Test failure.
Repair is a garage job.

Rubber-only suspension

These suspensions consist of suitably shaped rubber units
acting as springs. One distinct advantage they have is that they
are almost maintenance free. Rubber is a very good
storekeeper of energy. Another characteristic of rubber is that

the energy it releases after deflection is considerably less than the energy originally imparted to it. A rubber suspension can therefore act both as a suspension spring and partly as a shock absorber. As a result lower-duty shock absorbers can be substituted for normal ones.

The rubber "spring" is positioned between the vehicle frame and the suspension top link. Sometimes, the height of a vehicle to which a rubber-only suspension is fitted can drop after some usage; usually it is the front end that is affected. Agents for the make of vehicle concerned are normally able to renew the suspension or make adjustments to restore the vehicle to its original height. A suspension which is noticeably down would have to be failed.

Torsion-bar suspension

Fig. 25 shows a torsion-bar suspension arrangement. The secret of how a steel bar is made to perform the function of a spring lies in the material used in making the bar. The bars are made to withstand high stresses and to resist fatigue. They are made from high-carbon steel, or, more recently, from alloy steels such as silico-manganese. They are mostly circular in section, but can be rectangular or square. I will discuss the more usual (circular) sort first, and return to the latter two types in a moment.

The torsion bar is placed lengthwise along the vehicle. Both ends are splined. (Fig. 2 shows splines in a different situation if you are not sure what they look like.) The *rear* end of the bar (towards the back of the vehicle) is splined to a short piece of bar at a right angle to it. The unsplined end of this short piece of bar is bolted to the vehicle frame. (This is done, incidentally, in such a way that it can when necessary be adjusted, either to reduce, or to increase, tension on the torsion bar.)

The front end of the torsion bar is splined to the (lower) track control arm in such a way that, as the road wheel moves up and down, the torsion bar is "twisted", and, because of its internal construction mentioned above, is able to "spring" back. The *front* anchorage of the torsion bar is such that, whilst the bar itself cannot move, it can be twisted unhindered. It thus *suspends* the car "elastically" under twist tension, ready to absorb and respond to bumps in the road.

Square or rectangular torsion bars are only different because they do not need to be splined; their ends simply lock into a correspondingly shaped hole. However, a splined connection is much stronger and less liable to wear; hence the popularity of circular-section torsion bars over the others.

The mounting points of this suspension must be scrutinised for corrosion. Adjustment bolts must be tight and there must be sound structure for at least 30cm all round the rear mounting point, except where design precludes that much. In that case, however much there is must be clearly sound. A torsion bar front mounting bush (inside which the bar can twist as described) can squeak in old age; the squeak mainly disappears when it is wet, and then reappears when it is dry. Only replacement can stop the squeak for good. This is a garage rather than a D-I-Y job. Also the bar must not be slack inside this bush. This can be checked if you lever the bar up with a suitable implement. Any play should be immediately obvious. Substantial looseness would mean MoT Test failure.

The tie bar connection to the track control arm at the front tends to get loose frequently on some types and cause clonking − so this needs to be checked for security. (The tie bar is a bar which connects the track control arm to the vehicle frame at an angle of about 45°, to stop it swaying fore-and-aft.)

This suspension may sag after long mileage. The problem may be curable by adjustment or replacement but these are skilled jobs for a garage.

Tyres and wheels

Tyre inspection should start with the valve. The valve rubber casing has to be checked for cuts or greatly softened or degraded rubber, as well as for insecure insertion in the wheel rim. A tester must advise you of any potentially dangerous problem, so that the spare wheel can be fitted until repair is made. The same would apply to a leaky valve. The way to know if a valve is leaking is to smear a dollop of saliva over the airway mouth. Provided your saliva covers the whole circle of the mouth, air escaping will cause a bubble to form at once. (Incidentally, the MoT Test does not cover your spare wheel.)

Both side walls of each tyre must be checked for cuts and bulges. Any cut deep enough to reach the body cords renders the tyre unserviceable; any appreciable abnormal bulge or swelling raised above the surface of the tyre (that is, over the tread area or over the side walls) has to be failed.

With regard to tread, any separation or lifting of the tread from the rest of the tyre has to be failed. Any cuts through the tread sufficiently deep to interfere with the internal carcass of the tyre must be failed. While the law allows "retreads" to be fitted to vehicles, it does not allow "recuts" to be fitted to cars or light vans.

There must be no bald (smooth) patches of tread. The present law says three-quarters of the tread must be at least 1.6mm deep all round the circumference. The remaining quarter could be below 1.6mm but it *must not* be completely bald, and it *must* show a sign of tread pattern. The three-quarters of unworn tread area must be a continuous band. That is, you cannot have the well-worn quarter in the middle of the tread width with good tread on either side; the worn area can only be either on one edge, or on both edges of the tread width. A tyre depth gauge from motor accessory stores may help you to measure your tread depth. Some tyres have built-in wear indicators within the tread grooves. When the tread has worn down to the level of the indicators, the tyre has finished its useful life.

The tyre must be seated well on the rim, and badly distorted or buckled rims must be replaced. Readers are advised to check the latest tyre laws at a local tyre replacement/repair bay where the proprietors are legally bound to have the up-to-date laws on display.

The legal niceties of tyre laws can mean a pass or a fail of the MoT Test. As regards tread depth, in my opinion, anyone who allows their tyres to reach the minimum condition is very foolish. The risks of a blow-out, and the potential results of an unnecessary skid resulting in an accident, far outweigh the hoped-for economy of trying to squeeze the last mile out of a tyre. Tyres are the only link between your vehicle and the road, and perhaps between your life and your death or the deaths of others.

Wheel bearings

Front wheels are usually mounted on taper roller bearings, and tapered roller bearings must be allowed room for expansion, otherwise they seize up when they become heated. It is this room for expansion that shows as an end play or end float on a front wheel. Excessive end play is not permissible. See page 53.

3

STEERING CHECKS: WHERE TO JACK IT UP

To check the steering top and bottom swivels, where you jack up your vehicle is critical. If you position your jack wrongly, any wear that may be there can be taken out by the jacking up. The correct jacking point depends on the design of the suspension. The diagrams in this chapter (and those referred to) represent the most popular suspension designs.

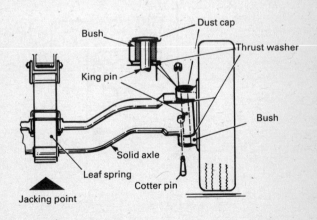

Fig. 26. Leaf-spring front suspension.

Fig. 26 is of a leaf-spring front suspension. The fat arrow indicates where you jack your vehicle up to examine for wear, at the king pin and bushes. The jack is best placed at the point the leaf spring is clamped to the axle. However, with this particular suspension arrangement, where to jack up is not as critical as for all other types. In practice, anywhere on the axle cross-beam is all right provided you don't raise the two front wheels together. That would be dangerously insecure.

Fig. 8 is of a wishbone and internal coil-spring type suspension. (The top and bottom arms look like a wishbone, hence the name.) Jack this suspension up at the position indicated by the arrow, in order to check your top and bottom swivel joints. A king pin and bushes may be fitted instead on some types. Your aim is to jack the suspension up as close to the lower swivel joint or king pin bottom bush as

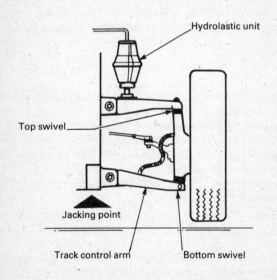

Fig. 27. Hydrolastic front suspension.

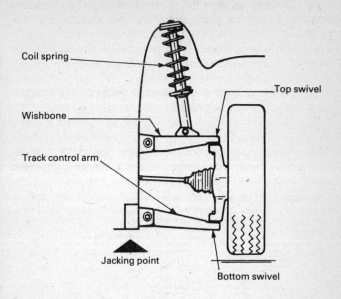

Fig. 28.   One type of coil-spring suspension.

possible. This is to ensure you compress the coil spring. If you don't place the jack underneath the spring, then once the road wheel is raised from the ground, the spring stretches and puts tension on the top and bottom wishbones. This, in turn, temporarily takes out any evidence of sloppiness in joints and swivels.

Thus, wear in swivel joints will be masked from you. With king pin and bush arrangements, any sideways wear will be equally hidden.

You must remember, therefore, always to jack up this type of suspension so as to compress the coil spring, but without jacking up the bottom ball joint itself (or the king pin), which of course is what you are checking.

Fig. 27 shows hydrolastic front suspension. Only jack this suspension up at the point arrowed. If you jack up the track control arm instead of a suitably strong part of the vehicle frame (or chassis), wear in the bottom ball joint is temporarily compressed out. (Although, as it happens, any wear in the top ball joint may still be detected.)

Fig. 28 is of another coil spring and wishbone combination

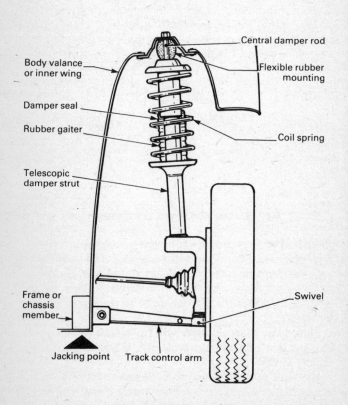

Fig. 29. McPherson-strut suspension.

— this time having the coil spring external to the wishbones. From the jacking up point of view it is the same as the hydrolastic suspension above. Hence my arrow to jack up the vehicle frame or chassis at a strong point.

Fig. 29 is of a McPherson-strut suspension. Jack up on the frame or chassis member as indicated by my arrow. On some McPherson-struts, wear in the bottom ball joint can be detected by jacking the vehicle up at this point, and rocking the wheel hard at the 12 o'clock and 6 o'clock positions. Unfortunately, on others, wear cannot always be detected this way, perhaps due to slight differences in design. I have therefore given four special tests for them under McPherson-strut suspension on page 101, to cover all eventualities. The track control arm must *never* be jacked up; if it is, any wear in the swivel joint will be compressed out by the track control arm pressing against the vertical strut.

4

CORROSION ASSESSMENT

In earlier chapters I have discussed superficial bodywork corrosion, hydraulic pipe corrosion and such matters. Here in this chapter, I am concerned with corrosion of load bearing body members, which, when sufficiently advanced, is going to be liable to affect the safety of the vehicle.

Underframe designs vary from the (usually older) classic chassis-based vehicle where the chassis does all the load bearing, with the body and suspension etc., being built onto it, to those where load bearing parts are simply integral panel members — but ones which have been strengthened by shaping. On the latter sort, the underfloor itself (around which these shaped members are created) must also generally be regarded as a load bearing structure. Sometimes under-bonnet flitch panels (e.g. inner wings) are load bearing too. Where front or rear suspension is mounted, or partly mounted, on them, you can be certain this is the case.

There are five basic underframe constructions currently in use. The first two are of the chassis-based construction already mentioned. See fig. 30 **A** and **A₁**. Here, independent chassis members run the whole length of the vehicle and do *all* the load bearing. In **A** they make a large rectangle and in **A₁** generally form an "**X**". They should be sound throughout. See later in this chapter how judgment should be made.

The next type of construction, in fig. 30 **B**, is by far the most common nowadays. In this, the front and rear sections of the underframe have load bearing box-sections, either pressed out or cast with the underframe, or welded to it during vehicle manufacture. Suspension, steering, etc., are naturally located onto these strengthened box-sections. In the middle of this underframe however, the only connection between the main box-sections, and between them and the outer sills which are load bearing too, may be plain floor. That floor therefore is also load bearing. All this load bearing structure must be sound − again, as will be described further into this chapter.

Fig. 30 **C** shows another method of construction essentially similar to **B**. Here, the whole vehicle underframe is manufactured with integral sub-frames. These sub-frames form each end and are extended to become the attachment for the door sills, with just plain floor supporting the middle. The load bearing structures are the sub-frames, the floor and the sills, and you must assess them all carefully for corrosion, on the lines suggested shortly.

The fifth construction type is shown at fig. 30 **D**. The central underframe, including the middle floor and some side panels, is pressed or cast out in one go. However, unlike the method of construction in fig. 30 **C** above, the front and rear sub-frames are *not* made an integral part; in this case they are bolted on later. One sub-frame is bolted to the front of this "basic box" to hold the engine and front suspension, the other to the back to hold the rear suspension. They are held on via tough, composite-rubber mountings (supported on each side by additional metal) which take up some road shocks. In addition to these main mountings, the two sub-frames are each held at their outer ends − again, via composite rubber mounts − to reinforced load bearing points incorporated in the vehicle bodywork. All the sub-frame mounting points are favourite areas for trouble, so inspect them extra carefully. Are fixings missing? Is a composite-rubber mounting degrading with age?

There are two toughened vertical panels (flitch panels) which run the width of the "basic box" and these are formed one at the front edge of it and one at the rear. It is onto these, unfortunately somewhat inaccessible, panels, that the respective sub-frames are usually held or partly held. The steering rack is also mounted on the front flitch panel.

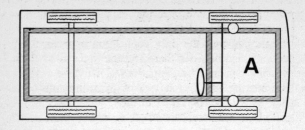

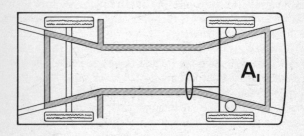

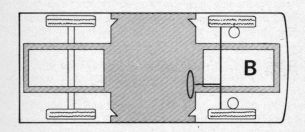

Fig. 30. (a) Underframes.

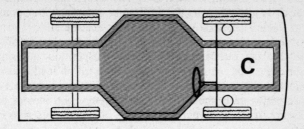

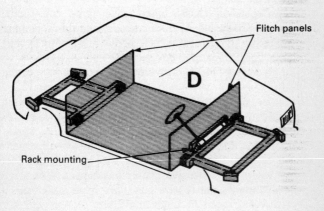

LOAD BEARING STRUCTURE

COMPOSITE RUBBER SUB-FRAME MOUNTING

Fig. 30. (b) Underframes.

The load bearing structures of this design which demand your close inspection are the sub-frames, their mounting points (four each), the flitch panels, the floor and the sills.

Step 9 of Chapter 1 calls for a thorough investigation of all load bearing structures. Since your suspension, steering, engine, gearbox, etc., all have to be mounted on load bearing members for obvious reasons, your investigations in respect of those major parts should give you a pretty good idea of the basic construction of your vehicle. Fig. 30 and the descriptions I have just given should be enough to confirm for you what type you have and where else you need to be looking.

You need to check the length and breadth of all the load bearing members.

The independent chassis-based underframe construction shown at fig. 30 **A** and **A₁** does corrode (and this sometimes happens before the general vehicle bodywork begins to rot!). Sub-frames, flitch panels, box-sections and underfloors are also highly susceptible to corrosion. *All* require very careful scrutiny.

Wherever you see surface rust that you suspect may be more extensive than meets the eye, it is worth tapping round the metal with the head of a medium screwdriver or similar implement, but don't poke. Severely rusted areas will return a softer thud than sound metal. This way, you should soon discover where any extensive rusting has happened or where improper fibreglassing repairs may have been attempted. A badly corroded metal or underframe box section will also squash or disintegrate when squeezed hard with the fingers.

You are looking for corrosion sufficiently advanced to weaken your vehicle's structure — to render a load bearing member no longer capable of so doing. MoT testers are told to look for loss of, or imminent loss of, *continuity* in load bearing members. Let that be your guide but, if in doubt, have a suspect area repaired anyway.

Corrosion in a load bearing structure must be expertly repaired in order to pass the MoT Test. You must therefore refer such a job to a specialist. Until recently, incidentally, all vehicle underframes were made of *metal* strong enough to withstand the stresses that an underframe is subjected to. However, nowadays some vehicle underframes are made of reinforced *plastic* able to withstand the same levels of stress. Again, repair work is for specialists only.

Remember that corrosion of *non* load bearing structures can also lead to MoT Test failure. For example, a rusted wing could be about to fall off. It would therefore be a potential danger to road safety and be classified as an Indirect Test Item. See Introduction.

5

SHORTCOMINGS OF THE MoT TEST

What is the MoT Test certificate worth?

In my considered opinion, the MoT Test certificate is worth very little; "very little" is, of course, better than nothing. The worth of a pass certificate depends on which garage carries out that Test. Some garages are very good and others are very bad. Nevertheless, the MoT Test holds the line against catastrophic standards of maintenance. What is needed now is for attempts to be made to improve it and for a greater safety consciousness to be spread amongst motorists.

Imagine this, not untypical, scenario. A motorist sends his car in for a Test, and collects it back with a pass certificate. He wonders why the tester overlooked the badly cracked flexible brake hose he had spotted earlier on, just before handing in the vehicle for the Test!

Grateful, however, that he has got a pass certificate anyway, he then sets out to renew the hose. On finishing the job, he finds that his brake pedal is not as firm as it used to be, despite his having bled the brake hydraulic system carefully afterwards as he had been told to do. Though puzzled as to why the pedal is not as hard as before, he consoles himself, saying "it is one of those things", and continues driving. He remains unaware that he hasn't done up one of the brake-hose unions tight enough, and that it is, therefore, leaking. Eventually and consequently, his brakes fail and he gets involved in an accident.

Who takes the blame? The bad workmanship is, of course, down to the motorist but, mind you, his car *had* just passed the MoT Test — and his brakes would probably soon have failed anyway …

Such sloppy testing clearly diminishes the value of the MoT Test. But inexpert D-I-Y work can, as illustrated above, redouble the problem. In this country anybody can dismantle his car and put it back together — at times lethally. Such motorists must therefore be as well-informed and safety conscious as possible. I hope my book will go a long way towards helping every driver become so!

Testers

Some time ago, I read of an investigative journalist who took the same car to about five different garages for an MoT Test. He got a failure certificate in each case. What amazed him most was that each garage put a *different* set of defects on its failure certificate!

Though testers are allowed slight discretion in passing judgment, I believe all five testers should have listed the same faults, if any, for the investigative journalist. An instance like this makes a mockery of the MoT Test.

The answer has to be that the quality of testers must be raised to a far higher level. Before someone becomes a tester, he ought to go on a far more rigorous course or undergo much better training than exists now.

It may surprise most people that for a mechanic to become a tester, the Department of Transport relies on the *garage to nominate* an experienced and intelligent mechanic. He then goes on a one-day course from about 9 a.m. to 4.30 p.m. Thereafter, if he fails to spot something dangerous when you hand over your vehicle to be tested, your life is, in a very real sense, in his hands. No *outside* additional qualifications — either practical or theoretical — are, at the time of writing, required.

Quality Inspection

Under Department of Transport (DoT) rules, garages having more than one tester must entitle one of them the *quality inspector*. His job in that regard is to double-check through about one out of every twenty tests carried out by each of his colleagues. He must overrule their decision if appropriate in the name of "quality".

What I find surprising is that, though the DoT sanctions the selection of the quality inspector, this inevitably gets done on the advice of the garage. Even more surprising is that garages with only one tester don't need to have a quality inspector! Why don't we have outside, independent inspectors? The quality inspector doesn't even need to be testing vehicles currently. He could be many years out of touch with the practicalities of testing. I refer to this again below. I hope the reader can see how garages can easily run rings around the system.

Test Station Inspectors

Having pointed out irrationalities in the system above, I should be fair to the DoT. Each year, a Department of Transport inspector visits every Test station to examine it and ensure that Tests are being carried out efficiently and honestly. At least this makes sense. Unfortunately, my observation is that most of these officials are more theoretical than practical, and they find it difficult to give practical advice to testers. Yet, if we are to improve the MoT Test, these officials ought to be the best-trained of all those involved in the testing business!

My suggestion is that the DoT should establish a sufficiency of their own direct Test stations carrying out MoT Tests for the public, to enable them to maintain a team of inspectors having hands on, up-to-date experience of testing. It is vital for them (as well as the Test stations they inspect) to have to carry out MoT Tests under workshop conditions (melting snow, etc., dripping on the tester) rather than just doing the occasional one under "clinical" conditions. Then their inspections and the advice they give to testers will become genuinely valuable. At the moment, the tendency is for the overall process to be bureaucratic, wasteful of time and unable to single out the rogues.

PART TWO

The purpose of this part of the book is not to turn everyone who reads it into a mechanic! It is mainly to show you how to carry out *simple* repairs arising from the Test. These are repairs most motorists should be able to do with a little common sense and sufficient enthusiasm. Practically every motor vehicle has a detailed workshop manual or equivalent written on it. I advise you to buy one to use in conjunction with this book if you wish to go beyond my scope here. I assure you it is rewarding and interesting to do so.

One secret of carrying out repairs successfully is not to panic if you start experiencing problems! For example, if you can't get a bolt through an awkward hole, or you have skinned your knuckles badly and are bleeding and it is getting dark or beginning to rain, leave the job for a while. Go inside for a cup of tea, and come back to continue with a restored sense of humour. Alternatively, arrange to continue the following day.

I give this homely advice because, once you reach an agitated frame of mind, your efficiency falls off. The task then takes longer and longer to do. Facing the work afresh makes you more efficient. However, if you can still make no progress, it's time to call in a professional.

I have repeatedly urged that you must always support your vehicle with proper equipment before you go under to work. Do not go under a jacked-up vehicle if it is only supported by a light-duty wheel-changing jack. Never risk precarious arrangements such as loose piles of bricks either. Safety must *always* be at the forefront of your mind — you are more important than the vehicle. Because we are anxious to get it done, it is easy, when a job is proving difficult to do, or we are running out of time, to ignore safety. Watch it!

Beginner mechanics frequently over-tighten nuts and strip threads, or even shear bolts. The best way to avoid this is to use a torque wrench. It is a special tool which can be set to the degree of tightness you want the bolt and/or nut. As you tighten and the wrench reaches this pre-determined torque setting, it "clicks" audibly and won't tighten any further. Though there are some bolts and nuts on a motor vehicle which you can tighten safely without a torque wrench, there are others for which the degree of tightening is critical. They require the use of a torque wrench at all times. Examples are cylinder-head bolts, tapered bearings, etc. Workshop manuals always specify torque settings for these important fixings.

However, if you don't have the manual and can't get hold of this tool, you will have to depend on your judgment. Naturally, the bigger the bolt/nut, the more torque (tightening) it requires, and vice versa. Part of the skill-training of mechanics is to learn how much tightening to put on a bolt/nut without under- or over-tightening, or stripping its threads. You may have to learn this from experience; however, a rough guide is to screw up until the nut or bolt is first tight — then just give it a half turn after that. Another way — which assumes you have the correct-size spanner and have not tried to make any extension to it (giving you excessive leverage) — is only to tighten until you begin to exert yourself slightly. Remember, if you do strip a thread or shear a bolt, particularly in an awkward place, you could well end up needing expensive professional help.

You can gain a great deal of mechanical experience preparing your vehicle for its MoT Test. Good luck!

6

A-Z OF D-I-Y REPAIRS

Quick Index to Subjects

Brakes

Extended brake creep test
Refer also to **Step 2** of Chapter 1.

1) Spring a piece of rod or a length of wood between the brake pedal and the bottom of the steering wheel, in such a way that light pressure is put on the brake pedal. Protect the steering wheel from this make-shift bar by means of a suitable soft pad.

2) Leave overnight.

3) A defective hydraulic system (e.g. a leaking brake pipe/hose, a leaking wheel cylinder, or a leaking master cylinder − or one which may not leak but will not hold pressure internally) will allow the brake pedal to creep slowly under this pressure. Eventually the piece of bar will fall out.

4) If the bar drops out overnight, you need to trace the source of the fault and have it put right.

Quick checks on a brake servo
If you have discovered during **Step 2** of Chapter 1 that the servo unit on your vehicle is probably faulty, first double-check that the brake master cylinder securing bolts/nuts are not loose. If they are, tighten them. Then try the test in **Step 2** again. If there are still problems, check that the vacuum hose that leads

from the servo unit to the engine inlet manifold (see fig. 1) is not holed, and that it is tight at both ends. If there is damage on this hose, undo the two clips, remove the hose and fit a new one. (It is a straightforward job. There is no "catch" to it.)

The next step, if the unit is still faulty, would be to replace the servo unit non-return valve. Move on to your workshop manual if you wish to tackle this yourself.

Disc-brake pads

There are many different types of disc brake but the basic principle is always the same. Hydraulic pressure squeezes disc pads on either side of a disc, onto that disc. The disc turns with the road wheel and is sandwiched between the (stationary) pads, ready for such braking effort to be applied. The system can be likened to the pads of a bicycle brake being squeezed onto the rim of the pedal cycle wheel. Fig. 10 shows a disc brake.

If you find that your brake pads need changing I must refer you to your garage or to working with a workshop manual. Every type has its own small differences with regard to changing the pads, too numerous for this book.

Incidentally, if ever you hear a grinding noise when you apply your brakes, go no further. It means that the frictional material on the pads is worn out and the metal parts of the pads are now grinding on the metal disc. Only if you catch it in time will you be able to get away without having to have new discs as well as new pads!

WARNING: If you do attempt to change pads yourself, *always pump the brake pedal several times afterwards.* This will bring the pedal up firm. Top up the fluid in the master cylinder too.

The reason for pumping up the pedal is that after brake pads' renewal, the first stroke on the pedal often takes it right to the floor! *So it is advisable not to let this happen on the street.*

After the pads' renewal, you may well need to *bleed* your brakes. However, you can renew pads without getting air into the braking system if you work with care. Again, follow your workshop manual or entrust the job to a good garage. Also, see *Brake bleeding*, below.

Disc brakes are self-adjusting. However, where a handbrake operates disc brakes, that may need some adjustment after new pads are installed, so check this out.

Brake shoes (in a drum brake)

While disc brakes have brake "pads", drum brakes have brake "shoes" − similarly lined with frictional braking material. The shoes are mounted inside a brake drum as in fig. 10. The drum turns with the wheel. The shoes, mounted on a stationary backplate, squeeze *outwards* against the circular wall of the drum when you brake.

As with disc brakes, the many different types preclude my giving detailed instructions here for changing brake shoes. You need a garage or a workshop manual and some mechanical aptitude!

After renewing brake shoes, remember the brakes will need *adjusting* (self-adjusters need to be brought up to their starting point too). The handbrake may need adjustment. The hydraulic system will probably require bleeding as well. Follow workshop manual instructions carefully, and, because a drum brake pedal also needs to be brought up firm, see my **warning** at the end of my disc brake notes, above. See also *Brake bleeding,* below.

Handbrake

If a handbrake releases itself or releases without your pressing the handbrake release button, this usually means the ratchet and pawl locking mechanism is faulty and it will have to be repaired or renewed. Fig. 31 shows the parts. They connect via a cable to actuating levers, usually at both rear wheels. Fig. 31 also shows those arrangements, at a typical rear drum brake.

If your handbrake operation is stiff, try to discover whether the stiffness is in the brake (in the disc or drum), in the cabling, or in the handle pivot. (The latter, more often than not, suffers corrosion.) To do so, first disconnect the cable actuating levers, on the back of the road wheel at each side. Then pull the handbrake/handle/lever up and down.

(**WARNING:** *the vehicle must be securely held against moving.*)

If the stiffness goes, then its source must lie within the brake assembly. However, if the stiffness persists, then it is either in the cabling or in the handle. Therefore now disconnect the

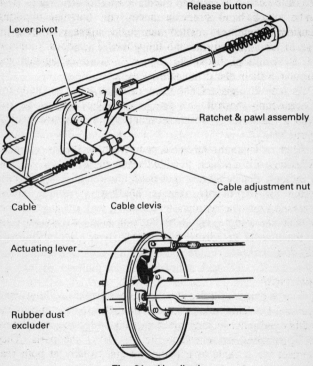

Fig. 31. Handbrake.

cable as close to the handle as possible. Again, work the handle up and down. If the stiffness disappears, then you know it is in the cabling. If it persists, then you have proved that it is in the handle. You can now investigate the precise cause accordingly, with the help of a workshop manual if necessary. Of course, there may be stiffness in two places at once; but your investigations ought to be able to track them both down!

Brake adjustment

On drum brakes, as the lining on the shoes wears down, an adjustment scheme must move the shoes nearer to the friction face of the drum in order to maintain braking efficiency. This can be done by hand at service intervals, or, more usually, self-adjusters (sometimes called automatic adjusters) maintain progressive adjustment as the lining wears away.

(Disc brake pads, by design, are in constant featherlight touch with their discs; hence they are self-adjusting anyway.)

When brake pedal travel becomes excessive you must always do something about it — a garage job unless you are expert yourself. However, it is worth saying that you should keep a mental note of how the brakes feel at the pedal when recently serviced, or when the vehicle is new. This gives you a personal benchmark from which to judge when things are wrong or servicing is clearly becoming overdue. Any sign of your brakes pulling the car to one side on application, or any other untoward sign, should always be investigated at once.

Brake servicing, pads' or shoes' replacement routines, etc. are not usually beyond the amateur but you need the proper workshop manual to guide you. Detecting and curing faults can be more difficult though manuals often list fault symptoms and appropriate cures. Remember that when a vehicle pulls to one side under braking, it is either weakness of the brakes on the *opposite* side of the car which will be the cause, or the brakes could be partially sticking on the side it pulls towards. See also *Brakes fault symptoms* below.

Brake bleeding

The hydraulic pressure which you, the driver, exert at the brake pedal is transmitted through the hydraulic fluid into the wheel cylinder(s) at each brake. See fig. 10. Within the cylinder(s) your pressure presses the piston inside up against the end of the brake shoe mounting or, in the case of disc brakes, the back of the pad mounting. In turn, that causes the friction lining material on the shoe or the pad, to be forced against the brake drum or brake disc accordingly. Hydraulic pressure is here neatly converted into mechanical force. Some manufacturers use similar hydraulics to operate clutches. The same principle is used to open and close flaps on aeroplane wings before take-off and during flight.

As you may recall from elementary physics a fluid cannot be squashed. Therefore your foot pressure transfers directly to the piston, assuming all is well. However, if air − a mixture of gases − gets into the brake fluid, it *can* be squashed. The result will be spongy braking at best or total failure at worst. You have to rid the system of air. The process is known as bleeding the brakes. The need to bleed them will theoretically only arise after some dismantling or repair work, or when air has entered the system via leaking wheel cylinders, leaking pipes, leaking unions (joints), etc. It may also happen because the brake master cylinder reservoir tide level has been allowed to fall too low. (This can happen because of sheer neglect or it can occur whilst linings are being replaced. However just removing and replacing brake shoes or pads should not usually necessitate bleeding the system unless a piston accidentally travels too far out of its cylinder in the process.) Whatever the cause may be, bleeding is normally the *last* job after the fault has been rectified.

Detailed procedures are to be found in workshop manuals but here are some general words of warning.

Never return fluid you have bled from the system, to top up the master cylinder. It is certain to be contaminated and/or aerated. Use fresh fluid of the correct type which has been standing for at least 24 hours. Before bleeding, remove any unnecessarily thick carpet or floor mat that may prevent the footbrake pedal from travelling its full distance during the bleeding process. Note: most brake fluid manufacturers recommend that the fluid in any system be changed anyway at least every two years.

Air shows itself in hydraulic fluid as bubbles. You will know air is eradicated when fluid being expelled ceases to have bubbles.

Throughout bleeding, the master cylinder reservoir must be topped up frequently; if the level falls too low, more air will simply enter the system. *Remember to put the top firmly back on when the job is complete.* Ideally, bleeding should be done by three people − one person at the road wheel, another topping up the master cylinder, and another pumping the pedal. All fluid must at all times be kept clean. Dust or debris must not get into it.

Sometimes bleeding will not dispel air from the system, even though an original fault allowing air into the system has been found and rectified. The problem may be caused by a backflow of air into a bleed nipple before it can be closed. If so, the procedure below (which is not in any text-book or workshop manual as far as I am aware, but which is very effective) can be tried:

1) Discard your bleed tube and the usual bottle already containing some fluid. But keep a glass jar, or other clear vessel, close at hand to catch expelled fluid.

2) Open the bleed nipple enough to give unrestricted passage of fluid, and, as brake fluid begins to drip, close it tight by placing your forefinger over the end.

3) Call to your assistant "pedal down". Timely removal of your forefinger will allow brake fluid to be ejected into your bleed jar whilst he gives the pedal a quick, strong, full downward stroke − but without air being allowed any chance of getting back in first.

4) After the ejection but *before* your assistant releases the pedal, you must close the nipple with your forefinger again (or you can tighten up the nipple with a spanner). You have to time it to close the nipple just before the flow would have stopped itself, and before any air could possibly flow back in.

Repeat this several times until you are satisfied all air has been eliminated. On the last occasion, make sure you tighten back the nipple *well towards the end of a firm downward pedal stroke,* and then replace the rubber cap.

Remember to keep the fluid reservoir topped up through the proceedings, always with fresh fluid which has had 24 hours' standing. (You will use more fluid than in normal bleeding.) Brake fluid "eats" paintwork. So wipe off any splashes immediately.

Brake fault symptoms

Whilst this section doesn't attempt to provide all the answers, more mechanically-minded readers may find useful pointers as to what is sometimes wrong. Naturally there can be other items too obscure to list.

1) Brakes cause vehicle to veer to one side

Harsh brakes on one side may be making the vehicle pull over that way. However, the brakes that side may be normal, with weak brakes on the other side being your problem. The trouble could stem from the back brakes, or the front ones.

Consider the following:

a) Piston in wheel cylinder seized up.

b) Shoe return spring broken.

c) Tyre treads markedly more worn on one side of the car than the other; tyre pressures uneven (tyres with the lesser pressure hold the road more); or types of tyre (e.g. radial ply, cross ply) wrongly mixed on one axle.

d) Disc loose on its hub, drum brake backplate fixings loose.

e) Wheel bearings badly worn.

f) Warped or severely chewed-up brake disc or drum.

g) Brakes on one side adjusted more tightly than on the other.

h) Suspension or steering fault, probably quite major.

i) Brake linings or pads on one side more efficient than the other, either because those on one side have become soaked in brake fluid or axle oil or hub grease, or because new brake linings have been fitted on one side and not the other (something which should NEVER be done — always replace on both sides at once).

j) Air in hydraulic system.

2) *Wetness around brake backplate*
   a) Wheel cylinders leaking.
   b) Axle oil leaking out due to unserviceable half-shaft oil seal.
   c) Hub grease leaking out due to unserviceable hub oil seal.

3) *Brakes grind on application*
   a) Disc pad lining worn through to metal backing, now grinding on disc.
   b) Brake shoe lining worn through to metal shoe, now grinding on drum.
   c) Foreign object (e.g. small stone) trapped in brake.

4) *Brakes judder*
   a) Loose bolts and nuts in suspension or in brakes.
   b) Badly chewed-up brake drum or disc.
   c) Distorted brake shoe or disc pad.
   d) Brake lining loose on shoe.
   e) Friction material loose on metal part of disc brake pad.
   f) Brake backplate loose on its mounting.
   g) Loose disc brake calipers.
   h) Loose wheel cylinder.

5) *Brakes squeak or squeal*
   a) Brake linings too polished (roughen up with emery cloth).
   b) Too much worn friction material (i.e. dust) trapped in drum.
   c) Dirt or road muck trapped in drum.

6) *Excessive brake pedal movement*
   a) Brakes need adjustment.
   b) Pedal fulcrum pin/bush worn, creating lost motion.
   c) Leak of hydraulic fluid out of the braking system.
   d) Wheel cylinders leaking.
   e) Brake master cylinder worn.
   f) Air in the hydraulic fluid.
   g) No brake fluid in reservoir.

7) Brake pedal feels "spongy"

Note: A servo-assisted brake pedal never feels "rock hard" and so may give the impression of "sponginess". Prior knowledge of how it should feel or comparison with a servo system known to be in good order, or expert opinion, needs to be your guide.

a) Main cause of this is air in hydraulic system.
b) Looseness of parts − e.g. of brake backplate, of lining on shoe, or of wheel cylinders on backplate, and so on.

Electrical circuits
(Earth faults, live connections, etc.)

Electrics' problems are perhaps the most feared sort on a motor vehicle; yet this should not be the case. The fear is caused by straightforward lack of understanding. The flow of electricity itself may be "mysterious", but problems usually only happen when it is interrupted. Therefore, the main thing you ever need to concentrate upon is how and why electrical connections go awry.

Start with the knowledge most of us appreciate from school, which is to accept that electricity always flows from a "live" side back to "earth", making a *circuit*. Then let me simplify the electrics on a motor vehicle.

For every electrical component − take a light bulb for example − a live wire is connected firmly to its live side from the live terminal (the positive side) of the battery. (Each electrical part has a live side and an earth side.) Another connection is made firmly from the earth side of the component back to the earth side of the battery (the negative side).

The circuit is in place all the time. It is simply "made" (completed) when you switch on the component, and "broken" (disconnected again) when you switch it off. See fig. 32. Between the battery and the unit you usually find that the wiring also breaks at *two* places, to install a fuse and a switch,

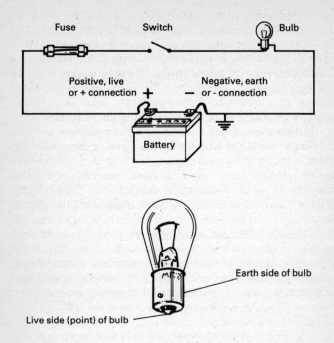

Fig. 32. Electric circuit, and bulb.

as shown.

These are the essentials of motor vehicle component circuits. When the circuit is also controlled by the ignition switch (i.e. only works when that is "ON"), that switch comes between the battery and the fuse.

If all the connections of all the electrical apparatus on a vehicle were made directly to the two battery posts, great sprays of wires would have to connect to them, and the cabling everywhere would look very ugly indeed. Therefore, in practice, one main connection is made from the battery + (or live) post to a point somewhere under the bonnet out of sight, controlled by the ignition switch where appropriate, and all the live connections are made from there, via the fuse box, to the live side of individual electrical units. Where several wires go in a similar direction most of their way, they are carried bundled together in what is called a loom, to keep things tidy. Individual wires are colour-coded and can be tracked using the wiring diagram for the vehicle from the workshop manual. Rather like underground railway maps, there is an art in reading these diagrams, soon learned by those interested.

Another way massive reductions in cabling are made is this. Instead of having to have earth wires leading from each unit back to the battery earth, negative (or −) post, *the metalwork of the vehicle is used*. One main earth connection is made by the battery earth post simply being bolted firmly to the vehicle's steel frame. The idea is that, since the vehicle structure is metal and there is continuity in it, electricity can travel through that instead of individual wires. Thus an earth connection made anywhere on the vehicle frame will conduct electricity back to the battery negative post, thereby completing that electrical circuit. For example, the earth connection from a stop lamp bulb is simply bolted to the nearest steelwork on the rear of the vehicle frame.

Nine times out of ten, when a unit isn't working and a known-to-be-sound replacement has been tried and still doesn't work, the fault will be found to be in that earthing path back to the battery.

Once you have checked that the fuse hasn't blown, trying a replacement is usually the first step, particularly with small item faults such as lamp bulbs, where, should the fault turn out not to be the bulb or whatever, having a spare in stock will be no bad thing anyway.

However, having a *test lamp*, obtainable from motorists' D-I-Y stores, is a huge advantage once you have to investigate beyond that sort of level. One is shown in fig. 33. It has a pointed probe to reach awkward areas. It also has a crocodile clip connection at the other end to make a firm grip to earth.

To see if current flow has reached an electrical part, first make sure the part is switched on! Then connect the earth clip of your test lamp firmly to somewhere on the main vehicle

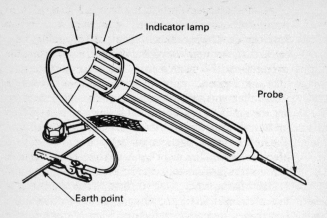

Fig. 33. Test lamp.

frame or the engine block, etc., where there can be no doubt
about the continuity of the earth passage back to the battery.
(The engine block is good because the engine always has an
earth connection back to the chassis or main frame. Ideally,
choose a place having bare metal if you are using the basic
vehicle structure.) Then probe the incoming live wire that
connects to the electrical unit, using the other end of your test
lamp. Your test lamp should light. If it doesn't, it means
electricity is not reaching the equipment.

When an electrical component stops working, your quickest
search route to pin-point the cause is likely to be as follows:

1)   Check the fuse covering that circuit.

2)   See that the connections either side of the unit are clean
      and tight, especially on the earth side.

3)   Use a test lamp (see above) to establish whether
      electricity is reaching the unit.

If electricity is NOT reaching the unit
− and the switch is on and the fuse is sound − there is probably what is called an "open" circuit. A break has happened somewhere along that incoming line, caused by the wire coming unclipped somewhere or maybe by its being severed. If you are an electrical Sherlock Holmes, you will be able to track back to joins in the wiring feed line, using the wiring diagram and the test lamp, until you find the section wherein the break must be. Repair may be as simple as cleaning up a bayonet-clip joint or rejoining a cut wire with a proper electrical joint. (These are cheaply available at motor accessory stores; never rely on taped-up bodging − that results in short-circuits and fires.) If a break is out of reach inside a loom, you may be able to add a substitute wire *outside* the loom. Make certain it is of the *same* electrical capacity as the original wire being replaced (i.e. wire of the same thickness and type, not thicker or thinner wire), and that the original wire is fully disconnected at both its ends and tidied away or removed.

If an open circuit hasn't occurred, the problem may be a *short-circuit*, or an intermittent short-circuit, wherein the electricity has found an "illegal" track direct home to earth from some point on the incoming feed line.

Short-circuits are dangerous because they heat up and cause fires. Perhaps the ultimate example would be were you to touch both battery poles at once with a metal spanner. Apart from a flurry of sparks, you would be likely to burn your hand. *Any short-circuit must be located and prevented from recurring*.

Lastly, you may find it is a new switch you need, or repair to one you have discovered is the cause of your problem.

If electricity IS reaching the unit
− next check the continuity of the earth return. You do

this by attaching your test lamp earth clip to the earth contact on the unit and taking the live-side probe across to the feed point on the unit, already known to be live. If the lamp lights, the component itself is probably faulty. If it *doesn't* light, the earth contact must be at fault, or there is poor contact somewhere between there and the vehicle main frame. A substitute temporary earth wire may prove the point.

4) Before condemning a component out of hand, always be certain the leads to and from it are making good contact with the item itself. For example, you may have ascertained that the feed to a bulb-holder is live and that the earth return is intact; but there may be corrosion in the bulb socket itself, sufficient to prevent the bulb from working. The bulb may light once you have cleaned out the socket thoroughly with fine sandpaper. A wire spade to bayonet clip attachment may appear tight but inside can be found a mass of verdigris. When that is sandpapered or wire-brushed clean to bare metal and fitted back together, hey presto! − the component works once more.

Lastly, on the subject of electrics, I stress that all electrical connections suffer if they are not *clean* and *tight*. This especially applies to the two battery post connections. Frequently, dirty post connections are all that is to blame when you cannot start your vehicle. A little maintenance lavished upon them from time to time is well worthwhile.

Remove the leads. Then clean off any acidic deposit on the posts with sandpaper. Smear the posts lightly with petroleum jelly before tightening the leads firmly back on.

WARNINGS: *Do not smoke.* Gases from a battery can explode. *Be extra careful not to short-circuit the battery;* see 3) above. *Protect* your hands and paintwork from any acidic material.

Exhaust repairs

Replacing the gasket at the cylinder head/manifold flange

In **Step 5** of Chapter 1 you may have discovered a leak here which you wish to repair yourself. (See also fig. 4.)

There is no catch in this job. On average, about eight nuts or bolts, four below and four above, secure the manifold to the cylinder head on a four-cylinder engine (twelve for a six-cylinder engine, and so on).

Remove all these nuts or bolts, taking a mental note of how tight they were. Carefully prise the manifold away from the cylinder head with a suitable lever, taking extreme care not to damage either the cylinder head, or the manifold. Watch that no undue strain is coming upon the manifold/downpipe flange or the exhaust pipe further out from there. If it does, or might, you will need to undo the manifold downpipe flange first, for which see below.

Remove the leaking gasket and meticulously scrape off any remaining vestiges thereof, *taking care not to allow gasket bits and pieces to enter the cylinder head or the manifold*. Install your new gasket *without* jointing compound. (Exhaust gaskets are fitted dry.) Ease the manifold back to mate with the cylinder head and re-tighten the bolts or nuts. If possible, use a torque wrench (see page 122) for this. Otherwise, take care not to over-tighten them. Use no more effort than you needed to undo them. Start engine and check if your leak has stopped.

Cracked exhaust manifolds can be welded if you can find a proper specialist to do it. This is a precision job, hence the need for expertise. If you cannot get it welded, then a cracked and leaking manifold has to be replaced to pass your MoT Test.

Repairing the joint at the manifold/downpipe flange

This flange may not always have a gasket. It may be a simple "ball and socket" type mating of the two parts, held together by clamps. There is sometimes a special sealing ring which fits between the parts — again the whole joint being clamped together. Alternatively there may be a "flat" joint, with bolts through and gasket between. This is usually put together (though not always) with a special jointing compound which can survive the heat.

The "ball and socket" (and sometimes sealing ring too) types occasionally leak because the clamps have just become loose. Tightening stops the leak. A new sealing ring may solve the problem. However, when an exhaust blow or a fracture has taken a bite out of the exhaust pipe around the mating flange, then that whole section of the pipe will need to be replaced. (Usually it is the pipe that holes, rather than the manifold, but the manifold can need replacement in a bad case.)

To mend a leaking gasket on the flat type of joint mentioned above, you need (unless it is of a dry-fitting sort) a supply of the right sort of jointing compound. This flange is held together by two, three or four bolts, or bolts-and-nuts. Remove the nuts or bolts securing the flange and remove the flange gasket. Thoroughly scrape off any remaining vestiges of the old gasket material. Install a new gasket − coated lightly with jointing compound (unless it's not specified) − and connect the two parts. Get someone to hold the downpipe firmly up against the manifold while you install and tighten the bolts/nuts.

Exhaust pipes and silencers

If an exhaust pipe is broken or perforated, you can either replace or repair (weld) it. If the general condition of the pipe is weak, you are best advised to replace the whole pipe at an exhaust centre. If only one section is bad, for example one containing a silencer box, they can usually renew just that section for you. (These centres do the work more quickly. You will only gain marginally if you buy to fix at home.)

However, if the pipe generally is in fairly good condition, you can bandage a small hole. Bandages are sold by accessory shops and they come with instructions. Essentially, you will be wrapping a bandage round a pipe and retaining it with steel wire. Such repairs are improvisations. It is left to the discretion of the tester whether to accept or reject them. So make sure you make a neat and tidy (and leak-free!) job, and don't attempt one on a pipe which is self-evidently about to collapse in several more places!

Once the external metal of a silencer box is holed, through ordinary wear, you can be fairly sure it needs replacement. The insides — the baffles — will be "shot", so there's no point in bandaging it.

Lighting

If a stop lamp, sidelamp, headlamp or an interior light, etc., has *failed*, turn to **Electrical circuits** earlier in this chapter. For direction indicator faults, see the next section. For headlamp *beam adjustment*, see directly after that.

Direction indicators

1) If a single direction indicator lamp is not coming on and flashing, the likelihood is that just that bulb has failed. The vehicle's instruction book will show how to fit a new one. If a known-to-be-good bulb doesn't work, make sure the bulb's socket is *clean*. If it still won't work but the other indicators all *do*, look for a faulty earth or feed wire to that socket. How to do so is explained under **Electrical circuits** a few pages back.

2) In **Step 3** of Chapter 1 I have explained the flashing rate requirements of indicators. If the rate appears to be too fast or too slow, suspect problems at one *or more* of the indicator lamps. (If a bulb has recently been replaced, did it have the correct voltage/wattage? The flasher unit and the bulbs are carefully matched in electrical rating to get the right "beat".) If all the indicators come on but don't flash, or they *all* flash at the wrong speed, and you are satisfied the bulbs *are* of the correct type, you probably need a new flasher unit. This is usually located under the dashboard. Note *exactly* how the wires are fitted to it before taking off the old one!

3) If *no* indicators come on or flash (remembering that your ignition needs to be on!), check the fuse first. If, after you replace the fuse, everything now works and it doesn't blow again at once, probably all is well. If it does blow, you need to find out where short-circuiting is causing this to happen. See **Electrical circuits** earlier in this chapter.

Assuming all is well with the fuse, you now need a test lamp (see page 135) to discover whether current is reaching the input wire on the flasher unit. You don't need to know which wire it is; just check through them all until you find it when the test lamp lights. Should none of them be alive, look for a fault between the fuse box and the flasher unit. Provided you find an input wire which is alive, you have narrowed matters down to there being a fault in the flasher unit *or* in the indicator switch. The likelihood, when *no* indicators come on or flash, is that it will be the flasher unit that's at fault. As it is usually simpler to substitute a known-to-be-working flasher unit than to fiddle with the indicator switch, that is best tried first. Don't risk short-circuiting a fresh flasher unit if a fuse originally blew. Find out the cause or you may "blow" the new flasher too.

If a new flasher unit doesn't solve your problem, turn to the indicator switch. If you are electrically adept, you may discover what's wrong. However, as these parts are often made *en bloc* − incorporating many other switches − you will probably have to resign yourself to having to purchase a new (and expensive) unit, and to having a vehicle electrician to fit it, unless, with luck, he can mend the original one.

4) If the indicators to one side of your vehicle work normally but those to the other side won't light up at all, check both apparently failed bulbs first. If known-

to-be-working bulbs substituted front and back don't solve your problem, check that "flashing" electricity is reaching those bulbs and/or that they earth properly before reaching any conclusion that you must replace the indicator switch itself.

Headlamp beam adjustment

Headlamp beams can be adjusted to focus higher or lower, or further outwards into the road or inwards towards the kerb. Each headlamp has an adjustment screw at the top or bottom for up and down adjustment, and another on one side for sideways alterations. Screw the spring-loaded adjuster in or out for the desired effect.

The proper aim can be checked against chalk marks drawn on a wall, as described below. Reasonably accurate settings can be obtained this way, if you work carefully, if ever adjustment should need to be made. The adjusters cannot usually be seen from the front of the vehicle; you must lift the bonnet and get to the rear of the lamp. On older cars, however, you may need to remove a chromed embellisher to the front of the lamp to reveal the adjustment screws.

You will need your vehicle parked facing *exactly* square (at right angles) to a straight vertical smooth wall on which you can chalk some position marks. The vehicle should be about one-and-a-half metres away from the wall. No fuss need be made about this distance; lights vary quite a lot and what you need is a reasonable gap that allows a good light pattern to be cast against the wall. However, it is important that the vehicle is standing on level ground. It is best if you can find a slightly dark place such as a lock-up garage; otherwise work at dawn or dusk. The greater light contrast available will enable you to identify the beam pattern(s) more clearly, and also to ensure that they are being cast properly in line with the length of the vehicle.

You will need an adult assistant to sit in the vehicle and switch the lights on/off and operate the dip switch as necessary. That person's weight also ensures the vehicle will be settled as if carrying a "normal" load; make sure you remove any "abnormal" loads such as gold bars, etc. from the boot!

Notes: If the on-the-move suspension of your vehicle is dependent on having the engine running, remember to have it ticking over when you set up your chalk marks (below) and for the tests. *Take care in this instance that you work in the open with no danger of being overcome by exhaust fumes.* If a manual adjustment is provided for temporary lowering of the lights when fully loaded, make sure it is set to the highest (i.e. normal load) position.

You now draw four chalk lines on your wall, ahead of the vehicle. See fig. 34. First, mark a horizontal line (parallel to the ground) at the same height off the ground as are your headlamp centres. Next, cut that line with a short vertical one which would match the exact lengthwise centre line of the vehicle were that to be extended out towards the wall. Finally, add two vertical lines to correspond exactly with your headlamp centres. As a check, you should find that these are placed

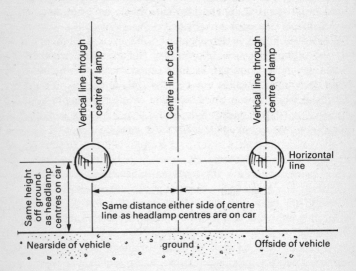

Fig. 34. Chalk lines on a wall, for headlamp beam adjustment.

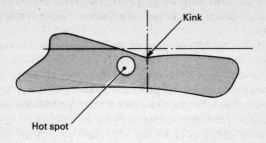

Fig. 35. European type headlamp beam adjustment.

equidistant on either side of the centre line you marked, just as your headlamp centres are from the centre line of the vehicle.

Refer back to **Headlamps** in Chapter 2, page 75, where the different types of headlamps are explained. You need *carefully* to identify from figs. 13 and 14 what sort you have got, and *on which beam they must be adjusted.*

Have your helper switch on your headlamps and set them on dipped or high beam as appropriate. But first note whether, when you move from high beam to dipped beam, the focus of light does move down, or down and slightly to the left. It must do this; otherwise you can take it you have the wrong type of headlamps, or perhaps headlamp bulbs, fitted — ones designed for driving on the opposite side of the road!

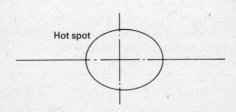

Fig. 36. British American type headlamp checked on *high* beam.

Cover one headlamp whilst you deal with the other one and vice versa. (Drape a small rug, or whatever, over it to black out the light.)

Remember that the purpose of the MoT headlamp beam test is not just to ensure you won't dazzle oncoming drivers; it is to ensure that your lights will maximise the vision you can obtain without causing that problem. So the regulations require that they mustn't aim too far down or too far leftward either.

For a European type headlamp (with the "staircase" pattern in the front glass) refer to fig. 35. Checking this on dipped beam, you should be able to identify a *hot spot* (having the most intense light). The whole of this hot spot must be in the south-west quarter, in relation to the centre point crossed by your lines on the wall in respect of this headlamp. The major area of light, which you should also be able to discern fairly precisely, is shown by fig. 35, with the hot spot inside it in the correct (south-west) quarter. The top edge of the major area of light ought to show a distinct kink as marked in fig. 35. That kink must play upon your *vertical* chalk line, and must be seen to cut it a little below your *horizontal* line. My picture shows the general positioning of these brightly lit areas in relation to your chalk lines.

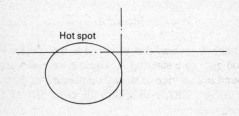

Fig. 37. British American type headlamp checked on *dipped* beam.

For British American types checked on *high* beam refer to fig. 36. Vertically, the *middle* of your hot spot should be bang on the horizontal cross line you have made on the wall in respect of the centre point of that headlamp. Horizontally, the *middle* should ideally be a small amount to the right of the vertical cross line you have made, as my picture shows. As long as it's pretty close to the middle you shouldn't have a problem.

For British American types checked on *dipped* beam refer to fig. 37. Vertically, the upper edge of your hot spot should encroach slightly over your horizontal line, as shown. Horizontally, the righthand edge of your hot spot should touch your vertical line but must not run over to the right of it.

Seat-belts

Seat-belts cannot be repaired. Manufacturers do not supply repair kits. Perhaps this is understandable considering the seriousness of the work a seat-belt does. Second-rate repairs could lead to death in an accident. If you find, in **Step 1**, Chapter 1, that your seat-belts are faulty, I'm afraid you will need to buy a new set to replace them. Fitting should be a straightforward loosening and tightening of bolts and nuts. Do remember to have a careful look at the bodywork all around the belt anchorages as emphasised in **Step 1.**

Steering

If you have discovered that you need to replace, for example, a track rod end, or a steering rack gaiter, itemised procedures can be found in a service or workshop manual for your vehicle. Such repairs are often well within the capability of a careful D-I-Y mechanic.

Remember that if you have power steering and you attempt, for instance, to replace a length of its pressure hose, you will then need to know how to bleed air out of that system in order to complete the job safely.

Wheel bearing adjustment to eliminate steering wander may be appropriate to your vehicle. Again, this can be a home repair job if you have the relevant information on how to do it to hand before you start.

Steering fault symptoms

This section may help you trace common faults. Generally, repairs will be garage jobs unless you are skilled and have the necessary workshop manual.

1) Steering wander

This is when the vehicle seems to be tracing somewhat of its own path even though it is being steered by the driver:

a) Steering joints worn or very worn.
b) Steering angles — castor, camber, etc. — badly altered through accident or forcibly knocking/mounting kerb.
c) Excessive wear in rack and pinion steering or of the gears in a steering gearbox.
d) Uneven tyre pressures or uneven tyre wear — including back tyres.
e) Stabiliser, or Panhard rod(s) loose.
f) Excessive end float on front wheel bearings.
g) Faulty suspension.
h) See 3) below if steering is also heavy.

2) Steering wobble

a) Wheels and tyres unbalanced.
b) Wheel rim distorted.
c) Examine a), b), c) and g) of steering wander, above.
d) Over-inflated tyre.
e) Failed shock absorber(s).

3) Stiff or heavy steering

a) Steering column partially seized.
b) Steering swivels stiff or partially seized.
c) Very low tyre pressures.
d) Damaged suspension, especially weak springs causing

the vehicle to drop in height or to be lop-sided.
e) Steering angles badly distorted, perhaps in an accident.
f) Steering damper very stiff in operation.
g) Incorrect toe-in or toe-out causing the tyres to scrub the road instead of rolling effortlessly.
h) No oil in rack and pinion or in steering gearbox.
i) Failure of servo-assistance.

Windscreen washer

A windscreen washer system consists of a (plastic) bottle to store the water/cleanser, (plastic) connecting tubes, a jet, or two jets, fitted at the base of the windscreen to direct pressurised water onto the windscreen, and a small pump (which is either manually or electrically operated).

If your washer stops working, it could be one of the following faults:

★ No water/cleanser in the reservoir.
★ Connecting tube holed, disconnected or damaged at some point on its route to the windscreen.
★ Pump not functioning or not functioning properly, or fuse in an electric pump circuit blown.
★ Jets blocked.
★ Air getting into the suction side of pump.
★ Water in the system frozen.

Let me go through the remedies:

Manual system
1) Make certain that the reservoir is not cracked or leaking, and that there is water/cleanser in it.

2) Ensure that a plastic tube connects firmly from the reservoir (with an air/fluid-tight connection) to the

suction (input) side of the pump. That tube should be full of fluid and clear of blockage. The plunger should be free for the driver to operate repeatedly, without leakage around it or undue stiffness.

3) Look to see that another (plastic) tube connects firmly from the pressure (output) side of the pump, to the water jets at the base of the screen, and that that tube is full and clear of blockage.

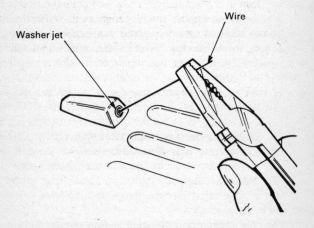

Fig. 38. Clearing a blocked windscreen washer jet.

4) If items 1) to 3) above are all right, then suspect the pump, but see 6) below: you may decide to poke the jets first before working on the pump.

5) Disconnect tube from output side of pump at its first junction *away* from the pump. Suck this tube to draw water/cleanser up through the pump. (It's better to suck directly on the pump outlet but you cannot always get at it with your mouth! You should notice at once if your problem is actually a hole in this piece of tubing or its

fitment onto the outlet.) If you can draw fluid, don't drink it (spit!); try the pump (briefly) again now. If it still doesn't spew fluid, the pump is faulty and needs replacement.

6) If, at 5), water ejects from the exit (output) side of the pump but the washer(s) still won't work when you put the piping back on, suspect blocked jets. Using a pair of pliers, pull out a strand from a wire brush and straighten it. Hold the strand firmly with the pliers, lean over your vehicle wing, and poke the fine holes of the jets to remove any blockage, as shown in fig. 38. Be careful that the wire does not break in the jet. If a stubborn blockage won't clear, remove jet and blow through it with an air line.

7) If ejected water aims above your windscreen or too low, bend the jet(s) carefully and gently to allow water/cleanser to hit the desired area of screen. Some jets are made of a small metal pipe with a pin-hole in it; these are bendable. Others consist of a ball with a pin-hole in it, placed in a socket. Use a strong, big needle to rotate the ball.

Electric pump system

The only difference between this and a manual system is the substitution of an electric pump for a hand one. So, apart from the pump, all the remedies I have outlined above apply to this system as well.

A switch, usually incorporated into the windscreen wiper control stalk but occasionally found on the dashboard, operates an electric pump/motor. This is either mounted on the water/cleanser reservoir or close to it; see fig. 39. This pump also has an input side and an output side. If you disconnect the output connection to the motor and switch on the washer, water should eject from this outlet if the pump is functioning. Also, these pumps usually operate noisily. So you will know the pump is faulty if there is no sound when you try the washers.

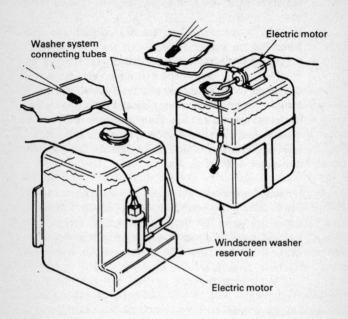

Fig. 39. Windscreen washer and pump.

When the problem seems to be a faulty motor, turn to my **Electrical circuits** section on page 133. You may find there is only an earthing fault, for example. If the motor does prove to be dud, replacement is usually quite easy. If the motor/pump functions, track out the problem just as for a *manual system* above.

In winter, the water/cleanser in the system may freeze. You *must* use one of the additives on the market to prevent the freezing. Apart from the danger of water splayed onto your screen turning instantly into ice and obliterating your view, it is illegal not to have a) a clean windscreen and b) windscreen wipers and washers in working order at all times. Not only that, an unsympathetic MoT tester may fail your vehicle if the water system is frozen!

Windscreen wiper fitting

A splined fitting most commonly attaches your wiper arm to the wiper motor power source. A spindle on the motor has male splines, which receive the female-splined end of the wiper arm. The latter is simply pushed on extremely firmly. See fig. 40. (Alternatively there may be a clamp system employing a pinch-bolt.)

To fit a new wiper arm, first lift the former one from the windscreen so that it stands upright on its own. Then, with a screwdriver or similar tool, prise the base of the arm upwards away from the motor spindle. Take care of the vehicle paintwork as you do so. (Obviously, a pinch-bolt type needs the bolt loosened off first.) To put on the new one, reverse the procedure.

Note that, depending upon the position in which you mate the splines, the wiper will, or will not, cover the screen fully. You need to get the right position so that the blade will not stop short in your line of vision; and neither will it slap the edges of the windscreen at either end of its travel. The best way to do this is to run the wiper motor whilst you observe the armless spindle. Stop it at the exact moment it reverses its travel. Then fit the new wiper arm and blade having pre-positioned it to match the corresponding edge of its travel across the windscreen.

Wiper arm and blade connections also have more than one type of fitting. The most common is the type shown in fig. 40. However, all wiper blades come with fitting instructions. Generally, look for a locating pin on the arm which snaps into a small hole on a swivel connection on the blade, and is held there by the tension of a small spring clip. Removal entails

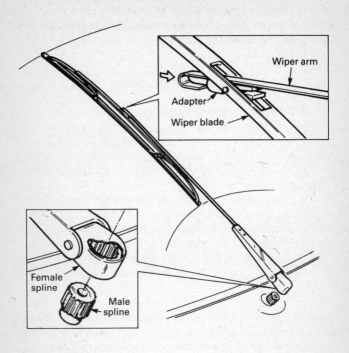

Fig. 40. Windscreen wiper arm and blade connections.

taking up the spring tension so that you can release the pin from its locating hole. Replacement is simply a matter of pushing it in until it snaps back in place.

WARNING: Do not attempt the above jobs while wearing a diamond studded ring. You may easily damage your windscreen by scratching it.

When a wiper blade is frozen to the windscreen (due to cold

weather) do not switch it on! The rubber blade is easily ripped apart this way. Instead, use de-icer to free the blade from the screen. Alternatively, scrape the ice off and gently lift the blade from the screen. If you hold the switch on and the blade cannot move, you can blow the fuse too.

INDEX

In the same series

CAR DRIVING IN TWO WEEKS

Over 1¼ million copies have been sold since this book was first written. Now in its 29th edition, it has been completely revised and updated to match the ever-changing conditions of our roads.

> *The Daily Telegraph:* "Immensely practical".
> *Birmingham Evening Post:* "Certainly the best".
> *Edinburgh Evening News:* "A notable contribution".

TEACH YOUR SON OR DAUGHTER TO DRIVE

A book for learner and teacher to use together. It consists of ten lessons centred around in-depth analysis of the Highway Code. It sorts out what to teach, in what order and *how*. Amateur and professional instructors alike will welcome the way David Hough illustrates the correct teaching principles that most quickly develop a pupil's competence.

HIGHWAY CODE QUESTIONS & ANSWERS

To pass your Test you *must* know the Highway Code thoroughly and be able to answer the specific questions on it which your examiner will ask. John Humphries' book contains 300 questions and answers designed to make learning the Code an easier task.

What To Watch Out for When BUYING A USED CAR

Thousands of people buy second-hand cars each year. Sadly, many of them end up with vehicles which cost hundreds of pounds in repair bills. Make sure you're not one of them. Find out how to judge a car's merits, detect its faults and value it accurately.

Whether you're buying from a private individual, from a car dealer or at auction, this book will save you money!

Uniform with this book

OUR PUBLISHING POLICY

HOW WE CHOOSE

Our policy is to consider every deserving manuscript and we can give special editorial help where an author is an authority on his subject but an inexperienced writer. We are rigorously selective in the choice of books we publish. We set the highest standards of editorial quality and accuracy. This means that a *Paperfront* is easy to understand and delightful to read. Where illustrations are necessary to convey points of detail, these are drawn up by a subject specialist artist from our panel.

HOW WE KEEP PRICES LOW

We aim for the big seller. This enables us to order enormous print runs and achieve the lowest price for you. Unfortunately, this means that you will not find in the *Paperfront* list any titles on obscure subjects of minority interest only. These could not be printed in large enough quantities to be sold for the low price at which we offer this series.

We sell almost all our *Paperfronts* at the same unit price. This saves a lot of fiddling about in our clerical departments and helps us to give you world-beating value. Under this system, the longer titles are offered at a price which we believe to be unmatched by any publisher in the world.

OUR DISTRIBUTION SYSTEM

Because of the competitive price, and the rapid turnover, *Paperfronts* are possibly the most profitable line a bookseller can handle. They are stocked by the best bookshops all over the world. It may be that your bookseller has run out of stock of a particular title. If so, he can order more from us at any time—we have a fine reputation for "same day" despatch, and we supply any order, however small (even a single copy), to any bookseller who has an account with us. We prefer you to buy from your bookseller, as this reminds him of the strong underlying public demand for *Paperfronts*. Members of the public who live in remote places, or who are housebound, or whose local bookseller is unco-operative, can order direct from us by post.

FREE

If you would like an up-to-date list of all *Paperfront* titles currently available, please send a stamped self-addressed envelope to
ELLIOT RIGHT WAY BOOKS, BRIGHTON ROAD.,
LOWER KINGSWOOD, SURREY, KT20 6TD, U.K.